D1546099

Integrals and Measures

PURE AND APPLIED MATHEMATICS

A Program of Monographs, Textbooks, and Lecture Notes

EXECUTIVE EDITORS—MONOGRAPHS, TEXTBOOKS, AND LECTURE NOTES

Earl J. Taft
Rutgers University
New Brunswick, New Jersey

Edwin Hewitt
University of Washington
Seattle, Washington

CHAIRMAN OF THE EDITORIAL BOARD

S. Kobayashi
University of California, Berkeley
Berkeley, California

EDITORIAL BOARD

Masanao Aoki
University of California, Los Angeles

Glen E. Bredon
Rutgers University

Sigurdur Helgason
Massachusetts Institute of Technology

G. Leitman
University of California, Berkeley

Marvin Marcus
University of California, Santa Barbara

W. S. Massey
Yale University

Irving Reiner
University of Illinois at Urbana-Champaign

Paul J. Sally, Jr.
University of Chicago

Jane Cronin Scanlon
Rutgers University

Martin Schechter
Yeshiva University

Julius L. Shaneson
Rutgers University

MONOGRAPHS AND TEXTBOOKS IN PURE AND APPLIED MATHEMATICS

1. *K. Yano,* Integral Formulas in Riemannian Geometry (1970)
2. *S. Kobayashi,* Hyperbolic Manifolds and Holomorphic Mappings (1970)
3. *V. S. Vladimirov,* Equations of Mathematical Physics (A. Jeffrey, editor; A. Littlewood, translator) (1970)
4. *B. N. Pshenichnyi,* Necessary Conditions for an Extremum (L. Neustadt, translation editor; K. Makowski, translator) (1971)
5. *L. Narici, E. Beckenstein, and G. Bachman,* Functional Analysis and Valuation Theory (1971)
6. *D. S. Passman,* Infinite Group Rings (1971)
7. *L. Dornhoff,* Group Representation Theory (in two parts). Part A: Ordinary Representation Theory. Part B: Modular Representation Theory (1971, 1972)
8. *W. Boothby and G. L. Weiss (eds.),* Symmetric Spaces: Short Courses Presented at Washington University (1972)
9. *Y. Matsushima,* Differentiable Manifolds (E. T. Kobayashi, translator) (1972)
10. *L. E. Ward, Jr.,* Topology: An Outline for a First Course (1972)
11. *A. Babakhanian,* Cohomological Methods in Group Theory (1972)
12. *R. Gilmer,* Multiplicative Ideal Theory (1972)
13. *J. Yeh,* Stochastic Processes and the Wiener Integral (1973) *out of print*
14. *J. Barros-Neto,* Introduction to the Theory of Distributions (1973) *out of print*
15. *R. Larsen,* Functional Analysis: An Introduction (1973)
16. *K. Yano and S. Ishihara,* Tangent and Cotangent Bundles: Differential Geometry (1973)
17. *C. Procesi,* Rings with Polynomial Identities (1973)
18. *R. Hermann,* Geometry, Physics, and Systems (1973)
19. *N. R. Wallach,* Harmonic Analysis on Homogeneous Spaces (1973)
20. *J. Dieudonné,* Introduction to the Theory of Formal Groups (1973)

21. *I. Vaisman,* Cohomology and Differential Forms (1973)
22. *B.-Y. Chen,* Geometry of Submanifolds (1973)
23. *M. Marcus,* Finite Dimensional Multilinear Algebra (in two parts) (1973, 1975)
24. *R. Larsen,* Banach Algebras: An Introduction (1973)
25. *R. O. Kujala and A. L. Vitter (eds.),* Value Distribution Theory: Part A; Part B. Deficit and Bezout Estimates by Wilhelm Stoll (1973)
26. *K. B. Stolarsky,* Algebraic Numbers and Diophantine Approximation (1974)
27. *A. R. Magid,* The Separable Galois Theory of Commutative Rings (1974)
28. *B. R. McDonald,* Finite Rings with Identity (1974)
29. *J. Satake,* Linear Algebra (S. Koh, T. Akiba, and S. Ihara, translators) (1975)
30. *J. S. Golan,* Localization of Noncommutative Rings (1975)
31. *G. Klambauer,* Mathematical Analysis (1975)
32. *M. K. Agoston,* Algebraic Topology: A First Course (1976)
33. *K. R. Goodearl,* Ring Theory: Nonsingular Rings and Modules (1976)
34. *L. E. Mansfield,* Linear Algebra with Geometric Applications (1976)
35. *N. J. Pullman,* Matrix Theory and its Applications: Selected Topics (1976)
36. *B. R. McDonald,* Geometric Algebra Over Local Rings (1976)
37. *C. W. Groetsch,* Generalized Inverses of Linear Operators: Representation and Approximation (1977)
38. *J. E. Kuczkowski and J. L. Gersting,* Abstract Algebra: A First Look (1977)
39. *C. O. Christenson and W. L. Voxman,* Aspects of Topology (1977)
40. *M. Nagata,* Field Theory (1977)
41. *R. L. Long,* Algebraic Number Theory (1977)
42. *W. F. Pfeffer,* Integrals and Measures (1977)

Integrals and Measures

Washek F. Pfeffer

Professor of Mathematics
University of California
Davis, California

MARCEL DEKKER, INC. New York and Basel

Library of Congress Cataloging in Publication Data

Pfeffer, Washek F
Integrals and measures.

(Monographs and textbooks in pure and applied mathematics ; 42)
Bibliography: p.
Includes index.
1. Integrals, Generalized. 2. Measure theory.
3. Riemann integral. I. Title.
QA312.P46 515'.42 76-29329
ISBN 0-8247-6530-3

COPYRIGHT © 1977 by MARCEL DEKKER, INC. ALL RIGHTS RESERVED

Neither this book nor any part may be reproduced or transmitted in any form or by any means, electronic or mechanical, including photocopying, microfilming, and recording, or by any information storage and retrieval system, without permission in writing from the publisher.

MARCEL DEKKER, INC.

270 Madison Avenue, New York, New York 10016

Current printing (last digit):
10 9 8 7 6 5 4 3 2 1

PRINTED IN THE UNITED STATES OF AMERICA

MATH-SCI
QA
312
.P46

TO LIDA,

who may not understand the book
but certainly understands the author

CONTENTS

PREFACE

The purpose of these notes is to develop the theory of measure and abstract Lebesgue integral by a method of extending certain linear functionals. This method, originally due to Daniell (see Ref. 7), seems to have some distinct advantages:

1. When applied to Euclidean spaces, it makes an actual use of the Riemann integral. The Lebesgue integral does not have to be built from scratch. Rather it is defined as a natural extension of the Riemann integral of continuous functions. Thus, the reader is not left with the unpleasant feeling that the Riemann integration which he learned previously has no value.
2. The proof of the Riesz representation theorem is essentially contained in the exposition. This, of course, is not surprising since virtually every elementary proof of the Riesz theorem is based on some modification of Daniell's method.
3. A systematic use of extension techniques can completely bypass the unintuitive Carathéodory definition of measurable sets.

Our presentation could be roughly described as a two-way bridge between integration and measure with the integral as a starting point. We begin with a certain system of functions, and we assume the integral of these functions has already been defined. For example, consider the Riemann integral of continuous functions on a compact interval. Next we extend this integral in three separate steps to a much larger family of functions, and using the extended integral, we define a measure. Proceeding in the other direction, we consider a measure and define an integral on a class of simple functions in a standard way. Extending this integral in a manner previously described, we obtain a new measure which turns out to be the Carathéodory extension of the original measure. The connection with the classical definition of an abstract Lebesgue integral follows immediately.

Throughout, the basic setting is an abstract space. Radon measures, i.e., integrals in topological spaces, emerge as a special case of the general contents. Therefore, a large part of the text is fully accessible to the reader who does not know any topology. At the same time a familiarity with basic properties of locally compact Hausdorff spaces will suffice for the understanding of almost everything (the exceptions are Chapter 18 and some exercises). Thus the book is suitable for the first quarter or semester of graduate lectures in real variables; before any topology is used, the student usually learns enough in his general topology course.

The first 13 chapters deal exclusively with integration over the whole space. They are closely related to each other and constitute the most important part of these notes. In Chapter 14, we define integrals over a measurable set and derive some of their basic properties. Chapters 15 to 17 are concerned with the products of measures. They present a nice application of the Daniell method developed in the previous chapters. In Chapter 18, we collect some relatively recent results on the regularity of Borel measures. Few of these results are new; others were previously scattered in the literature, sometimes in a hidden, not easily accessible way.

Since these notes contain mostly well-known material, they naturally overlap with other books in many places. I would like, however, to mention that my viewpoints on integration were largely influenced by the work of my teacher, Professor Jan Mařík (see Ref. 18). In particular, I learned a large part of the Lebesgue theory from the lectures of Professors Ilja Černý and Jan Mařík given at the Charles University in Prague (see Ref. 6).

The exercises collected at the end of each chapter contain many standard examples. Since the authorship of these examples is difficult to trace, I have made no effort to specify where they came from.

Davis, California W. F. P.

ACKNOWLEDGMENTS

It is a pleasure to acknowledge the help I obtained from many of my colleagues and friends.

Dennis Gannon and Gary Gruenhage read the original manuscript and made many valuable comments. The final version of the manuscript was carefully read by Steve Teel to whom I am obliged for a multitude of corrections and improvements. Saad Cherkaoui, Omar Hijab, and Larry Jones pointed out several mistakes and inaccuracies in the exercises. During the preparation of the manuscript I largely benefited from discussions with Carlos Borges, Don Chakerian, Magnus Giertz, Gary Gruenhage, Heikki Junnila, Mel Krom, Ben Roth, and Dennis Sentilles. In particular, Gary Gruenhage proved Theorem (18.11) which is published here with his permission.

The final chapters of this book were written in the Spring of 1975 while I was on a sabbatical leave from the University of California and visiting at the Kungliga Tekniska Högskolan (Royal Institute of Technology) in Stockholm. I am indebted to Magnus Giertz for inviting me to the KTH where I found a peaceful and stimulating atmosphere for my work.

Finally, my thanks belong to Tamsen BeMiller for typing the book so well.

W. F. P.

Integrals and Measures

0. INTRODUCTION

Unlike the preface, the introduction is aimed at a person who knows nothing about the subject and who intends to learn it from this book. It consists of two parts:

(a) A heuristic motivation for the development of the Lebesgue theory of integration.

(b) Some practical suggestions for reading this text; in particular, a careful list of the necessary prerequisites.

Heuristic motivation. Someone who spent a considerable amount of time and effort to master the Riemann integral, naturally, is not overjoyed at the prospect that there is yet another integral to learn. Therefore, a little advertising for the new integral is much in place. We shall do it in two ways: first, we shall show some serious deficiencies of the Riemann theory, and then we shall indicate how they can be corrected. It will follow that we are not abandoning the Riemann integral but merely extending it.

We shall begin with two examples.

(0.1) Example. Let c be a real number and let n be a positive integer. Define a function ϕ_n as follows:

$$\phi_n(x) = \begin{cases} n(x - c) + 1 & \text{if } c - \frac{1}{n} \leq x \leq c \\ -n(x - c) + 1 & \text{if } c \leq x \leq c + \frac{1}{n} \\ 0 & \text{if } x \leq c - \frac{1}{n} \text{ or } x \geq c + \frac{1}{n} \end{cases}$$

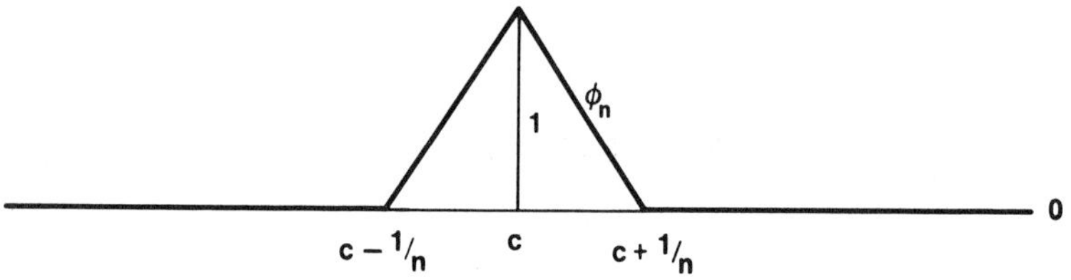

FIG. 0.1.

It is obvious that $\phi_{n+1}(x) \le \phi_n(x)$, n = 1, 2, ..., and that $\lim_{n\to+\infty} \phi_n(x) = \varphi_c(x)$ where

$$\varphi_c(x) = \begin{cases} 1 & \text{if } x = c \\ 0 & \text{if } x \neq c \end{cases}$$

Moreover, the Riemann integrals $\int_0^1 \phi_n(x)\,dx$ and $\int_0^1 \varphi_c(x)\,dx$ exist, $\int_0^1 \phi_n(x)\,dx \le 1/n$, n = 1, 2, ..., and $\int_0^1 \varphi_c(x)\,dx = 0$. Thus we have the following satisfactory situation: $\{\phi_n\}$ is a decreasing sequence of integrable functions, $\lim_{n\to+\infty} \phi_n = \varphi_c$, φ_c is an integrable function, and

$$\lim_{n\to+\infty} \int_0^1 \phi_n(x)\,dx = \int_0^1 \varphi_c(x)\,dx$$

(0.2) Example. Since the set of all rational numbers is countable, we can order it into a sequence, say $r_1, r_2, \ldots$. If x is a real number, let

$$\psi_n(x) = \begin{cases} 1 & \text{if } x = r_1, \ldots, r_n \\ 0 & \text{otherwise} \end{cases}$$

n = 1, 2, ... Clearly, $\psi_n(x) \le \psi_{n+1}(x)$ and $\lim_{n\to+\infty} \psi_n(x) = \psi(x)$, where

$$\psi(x) = \begin{cases} 1 & \text{if } x \text{ is rational} \\ 0 & \text{if } x \text{ is irrational} \end{cases}$$

Hence ψ is the well-known Dirichlet function which is not Riemann integrable. On the other hand, since $\psi_n = \varphi_{r_1} + \cdots + \varphi_{r_n}$, the functions ψ_n are Riemann integrable, and $\int_0^1 \psi_n(x)\,dx = 0$, $n = 1, 2, \ldots$. Thus, contrary to the previous example we have obtained the following highly <u>unsatisfactory</u> situation: $\{\psi_n\}$ is an increasing sequence of integrable functions, $\lim_{n\to+\infty} \psi_n = \psi$, $\lim_{n\to+\infty} \int_0^1 \psi_n(x)\,dx = 0$, and yet $\int_0^1 \psi(x)\,dx$ does not exist.

<u>Note</u>: The sequence $\{\phi_n\}$ from Example (0.1) is decreasing and the sequence $\{\psi_n\}$ from Example (0.2) is increasing. This fact is, however, quite inessential, as we can replace ψ_n by $1 - \psi_n$, $n = 1, 2, \ldots$.

The next theorem is the best result obtainable for the Riemann integral (see, for example, Ref. 1, Thm. 13-17, p. 405).

(<u>0.3</u>) <u>Theorem</u>. Let f and f_n, $n = 1, 2, \ldots$, be functions defined on an interval $[a,b]$. Suppose that

(i) $\int_a^b f_n(x)\,dx$ exists for each $n = 1, 2, \ldots$

(ii) $\lim_{n\to+\infty} f_n(x) = f(x)$ for each $x \in [a,b]$

(iii) There is a constant M such that $|f_n(x)| \le M$ for each $n = 1, 2, \ldots$ and each $x \in [a,b]$.

Then $\lim_{n\to+\infty} \int_a^b f_n(x)\,dx$ exists. If, in addition, $\int_a^b f(x)\,dx$ exists, then

$$\lim_{n\to+\infty} \int_a^b f_n(x)\,dx = \int_a^b f(x)\,dx$$

This theorem clearly demonstrates the strange state of affairs. While conditions (i)-(iii) guarantee the existence of $\lim_{n\to+\infty} \int_a^b f_n(x)\,dx$, they imply nothing about the existence of the integral $\int_a^b f(x)\,dx$. As Example (0.2) shows, this last integral may, indeed, not exist. To make things worse, one can actually prove that $\lim_{n\to+\infty} \int_a^b f_n(x)\,dx$ depends only on the function f and not on the choice of a sequence $\{f_n\}$ converging to f. More precisely, if $\{g_n\}$ and $\{h_n\}$ are sequences of functions on $[a,b]$ satisfying conditions (i)-(iii) of Theorem (0.3), then

$$\lim_{n\to+\infty} \int_a^b g_n(x)\,dx = \lim_{n\to+\infty} \int_a^b h_n(x)\,dx$$

In view of this, there is hardly any excuse for the nonexistence of $\int_a^b f(x)\,dx$. In other words, if the integral $\int_a^b f(x)\,dx$ does not exist, it is not because nature was unkind, but rather because the Riemann definition of the integral is too narrow. There is no reason whatsoever why we should not extend it and define the integral $\int_a^b f(x)\,dx$ by the equation

$$\int_a^b f(x)\,dx = \lim_{n\to+\infty} \int_a^b f_n(x)\,dx$$

(remember that the right side depends only on the function f and not on the sequence $\{f_n\}$). Carrying this extension out in a systematic manner is one of many ways in which the Lebesgue integral can be defined. We shall adopt this way and we shall present the abstract description of the whole extension process.

One more fact about Theorem (0.3) is noteworthy. While it follows quite easily from the Lebesgue theory [see (5.17) and (5-13) (iii)], its direct proof is excessively difficult. It is easy to believe that the integral which does not have the right properties and whose properties are hard to establish is not very useful on its own. Nonetheless, we shall use it as a basic building block for the Lebesgue theory.

Practical suggestions. We say right away that this text is not self-contained. In addition to some mathematical sophistication, the reader also should have a working knowledge in certain areas of advanced calculus, point-set topology, and "naive" set theory. On the other hand, there is no need for alarm as mostly only the rudiments are needed.

According to the necessary prerequisites, there are three levels at which this book can be studied.

(i) Minimal level. The reader should be familiar with basic properties of real numbers, elementary set operations, countable sets, topology of the real line, functions of one real variable, continuity, convergence, and the Riemann integral of continuous functions. With this knowledge he can read Chapters 1, 2, 4-8, 10-12, and 14 subject to the following modifications:

In Chapter 1, omit Section B and the discussion of the cardinals and ordinals from Section A.

In Chapter 4, formulate and prove Theorem (4.7) for the real line only.

In Chapter 10, omit Theorem (10.19).

In all chapters, omit the exercises involving topological spaces, cardinals, and ordinals.

Strictly speaking, Chapters 15 and 16 do not require any a priori knowledge of multiple integrals, however, they might be difficult to understand unless the reader knows some basics about the higher dimensional integration.

Any standard course of advanced calculus will certainly provide all prerequisites for the minimal level (for example, Ref. 1, Chapters 1-4, Chapter 9, Sections 1-15, and for the multiple integrals also Chapter 10, Sections 1-6).

(ii) Intermediate level. In addition to the advanced calculus topics listed in (i), the reader should know the theory of compact and locally compact Hausdorff spaces, their separation properties, and their finite products. With this knowledge he can read Chapters 1-17 but must omit the exercises involving paracompactness, cardinals, and ordinals.

Usually the first quarter or semester of a beginning graduate course in general topology will be an adequate prerequisite for the intermediate level (see, e.g., Ref. 16, Chapters 1, 3, 4, and the first four sections of Chapter 5).

(iii) Maximal level. In addition to the topics listed in (i) and (ii), the reader should be somewhat familiar with paracompact spaces, cardinals, ordinals, transfinite induction, and Zorn's lemma. Then he can read the whole book and do all of the exercises.

A good undergraduate course in "naive," i.e., nonaxiomatic, set theory will provide a sufficient background in cardinals and ordinals including transfinite induction and Zorn's lemma. A short and clear exposition is given, e.g., in Ref. 11 or, in a more condensed form, in Ref. 8, Chapter II.

All topological results we shall use can be found in Ref. 8, particularly in Chapters VII, VIII, and XI. We shall give exact references whenever properties of paracompact spaces are used.

Following our suggestions, the reader may proceed on each level without fear of disturbing the continuity of the exposition. With a few minor exceptions in Chapter 18, the main text is completely independent of the exercises. However, on each level there are many exercises which contain important material not covered in the main text.

The entire text depends in one way or another on Chapter 1. However, the later chapters may be grouped into several self-contained units. For the reader's convenience we are including a chart which indicates how the chapters depend on each other. This chart concerns the main text only and not the exercises.

CHAPTER DEPENDENCE CHART

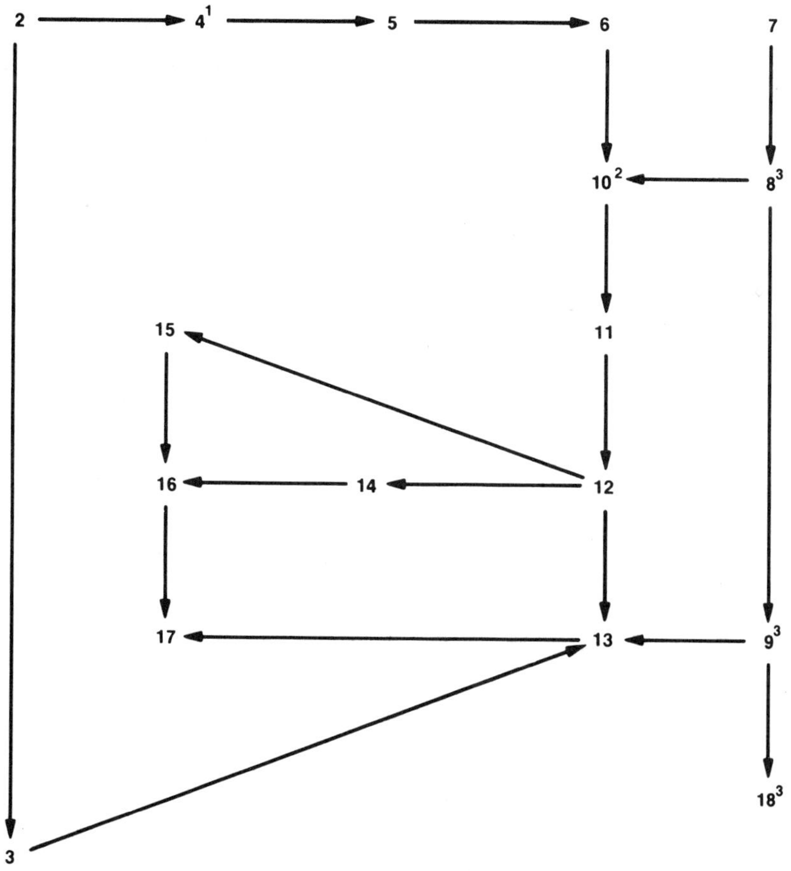

[1] Theorem (4.7) depends on Chapter 3.

[2] Theorem (10.19) depends on Chapter 9.

[3] Depends only on that part of Chapter 7 which precedes Definition (7.7)

1. PRELIMINARIES

In this chapter we shall establish the basic notation and terminology that will be used throughout the book.

(A) Sets. In our dealing with sets we shall always take the "naive" point of view. Thus a set will be a collection of arbitrary objects viewed as one entity. This "quasidefinition" of a set will be quite sufficient for our purposes; nowhere shall we attempt to use an axiomatic approach.

Standard notation of set theory, for example, $x \in A$, $A \subset B$, $A \cup B$, $A \cap B$, $A - B$, will be used freely without any explanation. The symbol $\emptyset$ denotes the empty set. The power set of a set A is the collection of all subsets of A, denoted by exp A. If C is a set and P is a property depending on $x \in C$, we shall denote by $\{x \in C : P(x)\}$ the set of all those $x \in C$ for which the property P(x) holds. Sometimes the set C is not specified and then we write only $\{x : P(x)\}$; for instance,

$$\exp A = \{B : B \subset A\}$$

The cartesian product of sets A and B is the collection of all ordered pairs (a,b), where $a \in A$ and $b \in B$; it is denoted by $A \times B$.

Let A be a set, and let $\{a_n\}$ be a sequence in A. Then $\{a_n\}$ is a map $n \mapsto a_n$ from the set of all positive integers to the set A. Sometimes, however, when no confusion can arise, we shall denote by $\{a_n\}$ also the set $\{a_n : n = 1, 2, \ldots\}$. In accordance with this convention we shall write $\{a_n\} \subset A$ to indicate that $\{a_n\}$ is a sequence in A.

A sequence $\{A_n\}$ of sets is called increasing if $A_1 \subset A_2 \subset \cdots$; it is called decreasing if $A_1 \supset A_2 \supset \cdots$. If $\{A_n\}$ is an increasing sequence of sets and $A = \bigcup_{n=1}^{\infty} A_n$, we shall write $A_n \nearrow A$. If $\{A_n\}$ is a decreasing sequence of sets and $A = \bigcap_{n=1}^{\infty} A_n$, we shall write $A_n \searrow A$.

A family $\mathfrak{C}$ of sets is said to be <u>directed upwards</u>, or <u>downwards</u>, if $\mathfrak{C} \neq \emptyset$ and for each pair of sets A and B from $\mathfrak{C}$ there is a set C in $\mathfrak{C}$ such that $A \cup B \subset C$, or $C \subset A \cap B$, respectively. If $\{A_n\}$ is an increasing, or decreasing, sequence of sets, then clearly, the family $\mathfrak{C} = \{A_n : n = 1, 2, \ldots\}$ is directed upwards, or downwards, respectively. If $\mathfrak{C}$ is directed upwards and $A = \bigcup \{B : B \in \mathfrak{C}\}$, we shall write $\mathfrak{C} \nearrow A$. If $\mathfrak{C}$ is directed downwards and $A = \bigcap \{B : B \in \mathfrak{C}\}$, we shall write $\mathfrak{C} \searrow A$.

If A is a set, then $\|A\|$ stands for the cardinality of A. We shall employ the usual notation

$$\aleph_0 < \aleph_1 < \cdots < \aleph_\alpha < \cdots$$

for <u>infinite cardinals</u> and

$$\omega_0 < \omega_1 < \cdots < \omega_\alpha < \cdots$$

for <u>infinite initial ordinals</u>. In this notation α is an arbitrary ordinal and ω_α is the <u>first</u> ordinal such that

$$\|\{\beta : \beta < \omega_\alpha\}\| = \aleph_\alpha$$

A set A is called <u>countable</u> if $\|A\| \leq \aleph_0$, and <u>uncountable</u> otherwise. Thus both finite and countably infinite sets are called countable. Because ω_0 and ω_1 play especially important roles among the initial ordinals, we shall denote them by ω and Ω, respectively. Clearly, ω is the first ordinal such that the set $\{\alpha : \alpha < \omega\}$ is infinite, and Ω is the first ordinal such that the set $\{\alpha : \alpha < \Omega\}$ is uncountable.

The cardinality of all real numbers is called the <u>continuum</u>, and it is denoted by c. If A is a set with $\|A\| = \aleph_0$, then $\|\exp A\| = c$. Of course, $c = \aleph_\alpha$ for some ordinal $\alpha \geq 1$. The <u>continuum hypothesis</u> asserts that $c = \aleph_1$. Whether it holds, cannot be decided, as either of the following assumptions is consistent with the axioms of set theory:

(i) $c = \aleph_1$

(ii) Given any ordinal $\alpha \geq 1$, $c > \aleph_\alpha$.

For the exact formulation and proof of this statement we refer the reader to Ref. 14, Chapters 12 and 18.

Unless specified otherwise, the continuum hypothesis will be assumed nowhere in this text. We shall denote by ζ the initial ordinal for c. Thus ζ is the first ordinal for which

$$\|\{\alpha : \alpha < \zeta\}\| = c$$

(B) Topology. Our topological terminology will adhere closely to that established in Ref. 16. As a rule, no properties of a topological space will be assumed tacitly. Thus, for example, unless specifically stated, compact, locally compact, and paracompact spaces are not necessarily Hausdorff, and σ-compact spaces are not necessarily locally compact.

If X is a topological space and A is a subset of X, we shall denote by A^- and A^o the closure and the interior of A, respectively. A subset A of a topological space X is called G_δ if there are open sets $G_n \subset X$, $n = 1, 2, \ldots$, such that $A = \bigcap_{n=1}^{\infty} G_n$; it is called F_σ if there are closed sets $F_n \subset X$, $n = 1, 2, \ldots$, such that $A = \bigcup_{n=1}^{\infty} F_n$.

If n is a positive integer, $\mathbf{R}^n$ will denote the n-dimensional Euclidean space with the ordinary topology. To us this will be, perhaps, the most important example of a locally compact Hausdorff space. Other important examples of locally compact and compact Hausdorff spaces will be, respectively, the sets $W = \{\alpha : \alpha < \Omega\}$ and $W^- = \{\alpha : \alpha \leq \Omega\}$ endowed with the order topology (see Ref. 16, Chapter 1, problem I, p. 57). While $\mathbf{R}^n$ is a metrizable space with many nice properties, the spaces W and W^- are nonmetrizable and, in fact, rather pathological. We shall use W and W^- mainly to produce counterexamples.

(C) Extended Real Numbers. By $\mathbf{R}$ we shall denote the topological ordered field of all real numbers. Of course, viewed as a topological space only $\mathbf{R} = \mathbf{R}^1$. The set $\mathbf{R}$ together with two ideal elements $+\infty$ and $-\infty$ is denoted by $\mathbf{R}^-$. The elements of $\mathbf{R}^-$ are called the extended real numbers. The order, the absolute value, and the algebraic operations in $\mathbf{R}^-$ are defined as follows:

$$-\infty < +\infty \quad \text{and} \quad -\infty < a < +\infty$$

for each $a \in \mathbf{R}$;

$$-(+\infty) = -\infty \qquad -(-\infty) = +\infty \qquad |+\infty| = |-\infty| = +\infty$$

$$a + \infty = +\infty + a = +\infty + \infty = +\infty \qquad a - \infty = -\infty + a = -\infty - \infty = -\infty$$

for each $a \in \mathbf{R}$;

$$a \cdot (+\infty) = (+\infty) \cdot a = +\infty \quad \text{and} \quad a \cdot (-\infty) = (-\infty) \cdot a = -\infty$$

for each $a \in \mathbf{R}^-$ for which $a > 0$;

$$a \cdot (+\infty) = (+\infty) \cdot a = -\infty \quad \text{and} \quad a \cdot (-\infty) = (-\infty) \cdot a = +\infty$$

for each $a \in \mathbf{R}^-$ for which $a < 0$;

$$0 \cdot a = a \cdot 0 = 0$$

for each $a \in \mathbf{R}^-$; $+\infty - \infty$, $-\infty + \infty$, $a/0$, $a/+\infty$, and $a/-\infty$ are not defined.

The reader should observe that with the exception of $0 \cdot (\pm\infty) = 0$, all algebraic operations in $\mathbf{R}^-$ are modeled along the well-known limit theorems for sequences of real numbers. We note that there is no "inner logic" in setting $0 \cdot (\pm\infty) = 0$. This rule is merely a convention which will prove useful for our purposes.

Operating with the extended real numbers is quite easy; nonetheless, some caution should be exercised. For example, let a and b be extended real numbers, and let $a \leq b$. Then we cannot automatically conclude that $b - a \geq 0$; for $b - a$ may have no meaning (e.g., if $a = b = +\infty$). If, however, we know that $b - a$ has meaning, then we can easily check that, indeed, $b - a \geq 0$.

The order in $\bar{\mathbf{R}}$ induces the order topology in $\bar{\mathbf{R}}$ which is obviously compatible with the usual topology of $\mathbf{R}$. The sets

$$(a,+\infty] = \{x \in \bar{\mathbf{R}} : x > a\} \qquad \text{and} \qquad [-\infty,a) = \{x \in \bar{\mathbf{R}} : x < a\}$$

form a local base at $+\infty$ and $-\infty$, respectively.

Every subset of $\bar{\mathbf{R}}$ has exactly one least upper bound and one greatest lower bound in $\bar{\mathbf{R}}$. Namely, if a set $A \subset \bar{\mathbf{R}}$ is not bounded from above or below in $\mathbf{R}$, then

$$\sup A = +\infty \qquad \text{or} \qquad \inf A = -\infty$$

respectively, and if $A = \emptyset$, then

$$\inf A = -\sup A = +\infty$$

For $a,b \in \bar{\mathbf{R}}$, we set

$$a \vee b = \sup\{a,b\} \qquad a \wedge b = \inf\{a,b\}$$

$$a^+ = a \vee 0 \qquad a^- = (-a) \vee 0$$

Clearly,

$$a = a^+ - a^- \qquad \text{and} \qquad |a| = a^+ + a^-$$

for each $a \in \bar{\mathbf{R}}$. If $a_i \in \bar{\mathbf{R}}$, $i = 1, 2, \ldots, n$, we let

$$\vee_{i=1}^{n} a_i = \sup\{a_1,\ldots,a_n\} \qquad \text{and} \qquad \wedge_{i=1}^{n} a_i = \inf\{a_1,\ldots,a_n\}$$

(D) Functions. By a function on a set A we shall always mean an extended real-valued function on A, i.e., a map $f : A \to \bar{\mathbf{R}}$. A real-valued function or a finite function on a set A is a map $f : A \to \mathbf{R}$.

The algebraic operations, partial ordering, and limits in the family of all functions on a set A are defined pointwise.

Let f and g be functions on a set A. If $f(x) + g(x)$ has meaning for each $x \in A$, we define the function $f + g$ on A by setting

$$(f + g)(x) = f(x) + g(x)$$

for each $x \in A$. The functions $f \cdot g$, $f \vee g$, $f \wedge g$, f^+, f^-, and $|f|$ are defined similarly. If

$$f(x) \leq g(x) \qquad \text{or} \qquad f(x) < g(x)$$

for each $x \in A$, we shall write

$$f \leq g \qquad \text{or} \qquad f < g$$

respectively.

Let A be a set, let f be a function on A, and let $\{f_n\}$ be a sequence of functions on A. If

$$\lim f_n(x) = f(x)$$

for each $x \in A$, we shall say that the sequence $\{f_n\}$ converges to f, and we shall write $\lim f_n = f$, or $f_n \to f$. Similarly, the symbols $\sup f_n$, $\inf f_n$, $\limsup f_n$, and $\liminf f_n$ are defined pointwise in the obvious way. The sequence $\{f_n\}$ is called increasing if $f_1 \leq f_2 \leq \cdots$, and decreasing if $f_1 \geq f_2 \geq \cdots$. If $f_n \to f$ and the sequence $\{f_n\}$ is increasing or decreasing, we shall write $f_n \nearrow f$ or $f_n \searrow f$, respectively.

Let A be a set, and let Φ be a family of functions on A. We shall say that the family Φ is directed upwards, or downwards, if $\Phi \neq \emptyset$ and for each pair of functions f and g from Φ there is a function h in Φ such that $f \vee g \leq h$, or $h \leq f \wedge g$, respectively. Let f be a function on A. If the family Φ is directed upwards and

$$f(x) = \sup\{g(x) : g \in \Phi\}$$

for each $x \in A$, we shall write $\Phi \nearrow f$. If the family Φ is directed downwards and

$$f(x) = \inf\{g(x) : g \in \Phi\}$$

for each $x \in A$, we shall write $\Phi \searrow f$. The symbol $\Phi \to f$ means that either $\Phi \nearrow f$ or $\Phi \searrow f$. If $\Phi \to f$, we shall call f the limit of Φ, and sometimes we shall write $f = \lim \Phi$.

If $\{f_n\}$ is a sequence of functions on A and $\Phi = \{f_n : n = 1, 2, \ldots\}$, then clearly, $f_n \nearrow f$ implies $\Phi \nearrow f$, and $f_n \searrow f$ implies $\Phi \searrow f$. Notice, however, that $f_n \to f$ does not imply $\Phi \to f$.

(E) Miscellaneous. With the exception of => and <=> denoting, respectively, an implication and an equivalence, we shall not use any logical symbols. We shall, however, write "iff" instead of "if and only if." Three slashes /// will indicate the end of a proof.

As is customary, most of the mathematical assertions stated in this text will occur under the headings of lemma, proposition, theorem, or corollary. These headings are used for emphasis, not to impose a more rigid formal structure on the exposition. Thus we want to encourage the reader to change them freely according to his own taste if he is inclined to do so.

For an easier orientation we shall use different types of numbering for the statements in the main text [e.g., (5.7)], for the figures (e.g., Fig. 3.1), and for the exercises [e.g., (7-8)]. The exercises which we consider hard are marked *, and those which we consider important are marked + . We would like to emphasize, however, that our considerations in marking the exercises may not agree with those of many readers. Thus, in particular, an asterisk by an exercise should not discourage the reader from attempting to solve it.

Exercises

(1-1)$^+$ Let $\aleph$ be an infinite cardinal, let T and A_t, $t \in T$, be arbitrary sets, and let $A = \bigcup \{A_t : t \in T\}$.

(i) In Ref. 11, Sec. 24, p. 97 it was proved that $\|B \times B\| = \|B\|$ for each infinite set B. Use this result to show that if $\|T\| \leq \aleph$ and $\|A_t\| \leq \aleph$ for each $t \in T$, then $\|A\| \leq \aleph$.

(ii) Show by example that in general $\|T\| < \aleph$ and $\|A_t\| < \aleph$ for each $t \in T$ do not imply that $\|A\| < \aleph$. Hint: Observe that $\{\alpha : \alpha < \omega_\omega\} = \bigcup_{n=1}^{\infty} \{\alpha : \alpha < \omega_n\}$.

(iii) Let α be the initial ordinal for $\aleph$, and let L be the set of all limit ordinals less than α. Show that if $\aleph > \aleph_0$, then $\|L\| = \aleph$. Hint: Observe that the map $\beta \mapsto \beta + \omega$ is a countable-to-one map from $\{\beta : \beta < \alpha\}$ into L, and apply (i).

Let $\{A_\alpha : \alpha \in Q\}$ be a nonempty family of sets. The cartesian product of the sets A_α is the collection of all maps $f : Q \to \bigcup\{A_\alpha : \alpha \in Q\}$ such that $f(\alpha) \in A_\alpha$ for each $\alpha \in Q$. It is denoted by $\Pi\{A_\alpha : \alpha \in Q\}$ or, if $A_\alpha = A$ for each $\alpha \in Q$, by A^Q. If $A_\alpha = \emptyset$ for some $\alpha \in Q$, then clearly $\Pi\{A_\alpha : \alpha \in Q\} = \emptyset$. On the other hand, it follows from the axiom of choice that $\Pi\{A_\alpha : \alpha \in Q\} \neq \emptyset$ if $A_\alpha \neq \emptyset$ for each $\alpha \in Q$. If $\beta \in Q$ we define the projection

$$\pi_\beta : \Pi\{A_\alpha : \alpha \in Q\} \to A_\beta$$

by letting $\pi_\beta(f) = f(\beta)$ for each $f \in \Pi\{A_\alpha : \alpha \in Q\}$.

(1-2)*+ Konig's theorem. Let Q be a nonempty set, and for each $a \in Q$, let A_α and B_α be sets such that $\|A_\alpha\| < \|B_\alpha\|$. Set $S = \bigcup\{A_\alpha : \alpha \in Q\}$, $P = \Pi\{B_\alpha : \alpha \in Q\}$. Prove that $\|S\| < \|P\|$.
Hint: Using the fact that any set of cardinals is well ordered, it suffices to show that there is no map $\varphi : S \to P$ which is onto. Suppose there is. Because $\|A_\alpha\| < \|B_\alpha\|$, we can choose an $f \in P$ so that $f(\alpha) \in B_\alpha - \pi_\alpha \circ \varphi(A_\alpha)$ for each $\alpha \in Q$. Since φ is onto, there is a $\beta \in Q$ and an $x \in A_\beta$ such that $\varphi(x) = f$. A contradiction follows.

(1-3)+ Let $A = \{0,1\}$, and let Q be a nonempty set. Show that

(i) $\|A^Q\| = \|\exp Q\|$. Hint: For $A \subset Q$, let $\chi_A(\alpha) = 1$ if $\alpha \in A$, and $\chi_A(\alpha) = 0$ if $\alpha \in Q - A$. Observe that $A \mapsto \chi_A$ is a one-to-one map from exp Q onto A^Q.

(ii) $\|Q\| < \|\exp Q\|$. Hint: Use (1-2).

(1-4)$^+$ Transfinite induction. Let α and β be ordinals, $\alpha < \beta$, and let $A \subset \{\gamma : \alpha \le \gamma < \beta\}$.

Suppose that

(a) $\alpha \in A$

(b) If γ is an ordinal, $\alpha < \gamma < \beta$, and if $\{\delta : \alpha \le \delta < \gamma\} \subset A$, then $\gamma \in A$.

Prove that $A = \{\gamma : \alpha \le \gamma < \beta\}$. Hint: Suppose $A \neq \{\gamma : \alpha \le \gamma < \beta\}$ and look at the first ordinal γ for which $\alpha \le \gamma < \beta$ and $\gamma \notin A$.

(1-5) Let $a, b \in \mathbf{R}^-$, $\{a_n\} \subset \mathbf{R}^-$, $\{b_n\} \subset \mathbf{R}^-$, and let

$$\lim a_n = a \qquad \lim b_n = b$$

Find conditions under which the following equations hold.

$$\lim(a_n + b_n) = a + b$$

$$\lim a_n b_n = ab \qquad \lim \frac{a_n}{b_n} = \frac{a}{b}$$

(1-6) Show that

(i) If a, b, and c belong to $[0, +\infty]$, then

$$(a + b) \wedge c = (a \wedge c + b \wedge c) \wedge c$$

(ii) If a, b, and c belong to $(-\infty, +\infty]$, then

$$(a - b) \wedge c = a \wedge (b + c) - b$$

whenever either side is defined.

(iii) If $a, b, c \in \mathbf{R}^-$ and $a \le b$, then

$$b - a \ge b \wedge c - a \wedge c$$

whenever both sides are defined.

(1-7) Show that $\mathbf{R}^-$ is a compact Hausdorff space.

(1-8)* For x and y in R, let

$$\rho(x,y) = |\arctan x - \arctan y|$$

where arctan x denotes the inverse function of tan x. Show that

(i) ρ is a metric in R which gives R its usual topology.

(ii) The metric space (R, ρ) is not complete. Hint: Show that $\rho(m,n) \to 0$ as $m,n \to +\infty$.

(iii) ρ can be extended to a metric ρ^- on R^- which gives R^- its usual topology.

(iv) The metric space (R^-, ρ^-) is the completion of (R, ρ) [see Ref. 13, (6.85), p. 77].

(1-9) Let $Q = \{1,2,\ldots\}$, and let $X = [0,1]^Q$. For x and y in X, let

$$\sigma(x,y) = \Sigma_{n=1}^{\infty} 2^{-n}|x(n) - y(n)|$$

and show that

(i) σ is a metric in X.

(ii) If $x \in X$ and $\{x_k\} \subset X$, then $x_k \to x$ in X iff $\lim_{k\to\infty} x_k(n) = x(n)$ for $n = 1, 2, \ldots$.

(iii) The space (X, σ) is separable. Hint: Let D consist of those $x \in X$ for which $x(n)$ is a rational number for all n and $x(n) = 0$ for all but finitely many n. Show that D is a countable dense subset of X.

(iv) $\|X\| = c$. Hint: Use (iii).

(1-10)+ Let $\{A_n\}$ be a sequence of sets such that $\|A_n\| < c$, $n = 1, 2, \ldots$. Prove that $\|\bigcup_{n=1}^{\infty} A_n\| < c$. Hint: Use (1-2) and (1-9)(iv).

Compare this result with (1-1)(ii).

(1-11) An ordinal α is said to be cofinal with ω if there is a countable sequence of ordinals $\alpha_1 < \alpha_2 < \cdots < \alpha$ such that $\lim \alpha_n = \alpha$. Show that if α is cofinal with ω, then $c \neq \aleph_\alpha$. Hint: Use (1-10).

(1-12)$^+$ Show that every metrizable space is perfectly normal, i.e., it is normal and each closed subset is G_δ (see Ref. 16, Chapter 4, problem K, p. 134).

(1-13) Let F be the set of all limit ordinals in W. Show that F is closed but not G_δ. Hint: Observe that if $G \subset W$ is open and $F \subset G$, then W - G is finite.

Using (1-12) conclude that the space W is not metrizable.

(1-14)$^+$ Let $\{x_n\}$ and $\{y_n\}$ be sequences in W such that $x_n \leq y_n \leq x_{n+1}$, n = 1, 2, ... Show that

$$\lim x_n = \lim y_n = z$$

for some $z \in W$.

(1-15)$^{*+}$ Let f be a continuous function on W. Show that

(i) If a and b belong to $\mathbf{R}^-$ and $a < b$, then one of the sets $\{x \in W : f(x) \leq a\}$ and $\{x \in W : f(x) \geq b\}$ is countable. Hint: Use (1-14).

(ii) There is an $a \in \mathbf{R}^-$ and an $x_0 \in W$ such that $f(x) = a$ for each $x \in W$ for which $x \geq x_0$. Hint: Use (i).

(1-16) Prove the following:

(i) W^- is the one-point compactification of W (see Ref. 16, Chapter 5, p. 150).

(ii) W^- is normal.

(iii) W^- is not first countable. Hint: Show that Ω has no countable neighborhood base.

2. FUNDAMENTAL SYSTEMS OF FUNCTIONS AND THEIR FIRST EXTENSIONS

As we explained in the Introduction (Chapter 0), we shall build the integration theory by a systematic extension of a given integral, e.g., the Riemann integral, to larger and larger families of functions. Naturally, this extension process has to begin somewhere. Since our aim is to capture the situation in its full generality, we shall use an abstract description of fundamental systems of functions on which integrals can be defined. A good, simple example to keep in mind for motivation purposes is the family of all continuous functions on a compact interval $[a,b]$.

(2.1) Definition. A fundamental system on a set X is a nonempty family $\mathfrak{F}$ of real-valued functions on X satisfying the following conditions:

(i) $f,g \in \mathfrak{F} \Rightarrow f + g \in \mathfrak{F}$

(ii) $f \in \mathfrak{F}$ and $a \in \mathbf{R} \Rightarrow af \in \mathfrak{F}$

(iii) $f \in \mathfrak{F} \Rightarrow |f| \in \mathfrak{F}$.

Throughout this chapter we shall assume that X is an arbitrary set and that $\mathfrak{F}$ is a fundamental system on X.

Conditions (i) and (ii) imply that $\mathfrak{F}$ is a vector space over $\mathbf{R}$. In particular, since $\mathfrak{F}$ is not empty, it follows from (ii) that the function identically equal to zero belongs to $\mathfrak{F}$. Because for real-valued functions

$$f \vee g = \frac{1}{2}(f + g + |f - g|) \qquad \text{and} \qquad f \wedge g = \frac{1}{2}(f + g - |f - g|)$$

we have

(2.2) $$f, g \in \mathfrak{F} \Rightarrow f \vee g,\ f \wedge g \in \mathfrak{F}$$

Thus, in particular,

(2.3) $$f \in \mathfrak{F} \Leftrightarrow f^+, f^- \in \mathfrak{F}$$

This last observation is quite useful as it frequently enables us to consider nonnegative functions only.

A vector space which satisfies condition (2.2) is called a vector lattice. Abstract vector lattices have many important properties which reach beyond the scope of this book. We refer the interested reader to Ref. 4, Chapter XV.

The extreme examples of a fundamental system $\mathfrak{F}$ are

(i) $\mathfrak{F}$ consists of all real-valued functions on X

(ii) $\mathfrak{F}$ consists only of the function identically equal to zero.

Both of these extreme cases are not very interesting.

We denote by $\mathfrak{F}_+^*$ the system of all functions f on X for which there is a sequence $\{f_n\} \subset \mathfrak{F}$ such that $f_n \nearrow f$. Analogously, by $\mathfrak{F}_-^*$ we denote the system of all functions f on X for which there is a sequence $\{f_n\} \subset \mathfrak{F}$ such that $f_n \searrow f$.

We obtain immediately

(2.4) $$\mathfrak{F} \subset \mathfrak{F}_+^* \cap \mathfrak{F}_-^*$$

(2.5) $$f \in \mathfrak{F}_+^* \Leftrightarrow -f \in \mathfrak{F}_-^*$$

(2.6) $$f \in \mathfrak{F}_+^* \Rightarrow f > -\infty$$
$$f \in \mathfrak{F}_-^* \Rightarrow f < +\infty$$

(2.7) If $f, g \in \mathfrak{F}_+^*$ and $a \in \mathbb{R}$, $a \geq 0$, then $f + g$, $f \vee g$, $f \wedge g$, and af belong to $\mathfrak{F}_+^*$.

(2.8) $$f \in \mathfrak{F}_+^* \Rightarrow f = \sup\{g : g \in \mathfrak{F},\ g \leq f\}$$

Let $\{f_n\} \subset \mathfrak{F}$ and $f = \sup f_n$. If $g_n = \vee_{k=1}^n f_k$, then by (2.2), $g_n \in \mathfrak{F}$ and, of course, $g_n \nearrow f$. Thus we have

(2.9) $f \in \mathfrak{F}_+^*$ iff there is an arbitrary (i.e., not necessarily increasing) sequence $\{f_n\} \subset \mathfrak{F}$ such that $\sup f_n = f$.

Moreover, $\mathfrak{F}_+^*$ is closed with respect to taking limits of increasing sequences. We shall formulate this precisely.

(2.10) Proposition. Let $\{f_n\} \subset \mathfrak{F}_+^*$ and $f_n \nearrow f$. Then there is a $\{g_n\} \subset \mathfrak{F}$ such that $g_n \leq f_n$, $n = 1, 2, \ldots$, and $g_n \nearrow f$. In particular, $f \in \mathfrak{F}_+^*$.

Proof. As is to be expected, the proof is a version of the "diagonal argument." Choose $\{f_{n,k}\}_{k=1}^\infty \subset \mathfrak{F}$, $n = 1, 2, \ldots$ such that $f_{n,k} \nearrow f_n$ as $k \to +\infty$, and let

$$g_n = \vee_{i,k=1}^n f_{i,k}$$

Then $g_n \leq g_{n+1}$, and by (2.2) $g_n \in \mathfrak{F}$. Since

$$f_{i,k} \leq f_i \leq f_n$$

for all $i \leq n$ and $k = 1, 2, \ldots$, we have $g_n \leq f_n$. Thus $g_n \nearrow g$ for some function $g \leq f$. On the other hand, $g_k \geq f_{i,k}$ for all $k \geq i$. Thus passing to the limit as $k \to +\infty$ we obtain $g \geq f_i$ for $i = 1, 2, \ldots$, and hence also $g \geq f$. ///

(2.11) Corollary. If $\{f_n\} \subset \mathfrak{F}_+^*$ and $f = \sup f_n$, then $f \in \mathfrak{F}_+^*$.

Systems $\mathfrak{F}_+^*$ and $\mathfrak{F}_-^*$ were obtained by adding to $\mathfrak{F}$ all limits of monotone sequences. We shall also need a larger extension of $\mathfrak{F}$ which is obtained by adding to $\mathfrak{F}$ all limits of directed systems. Recall that directed systems of functions and their limits were defined in Chapter 1, Section D.

We denote by $\mathfrak{F}_+^\#$ and $\mathfrak{F}_-^\#$ the system of all functions f on X for which there is a system $\Phi \subset \mathfrak{F}$ such that $\Phi \nearrow f$ and $\Phi \searrow f$, respectively.

Using the definitions only, we obtain results analogous to (2.4) through (2.7).

(2.12) $$\mathfrak{F} \subset \mathfrak{F}_+^\# \cap \mathfrak{F}_-^\#$$

(2.13) $$f \in \mathfrak{F}_+^\# \Longleftrightarrow -f \in \mathfrak{F}_-^\#$$

(2.14) $$f \in \mathfrak{F}_+^\# \Rightarrow f > -\infty$$
$$f \in \mathfrak{F}_-^\# \Rightarrow f < +\infty$$

(2.15) If $f, g \in \mathfrak{F}_+^\#$ and $a \in \mathbf{R}$, $a \geq 0$, then $f + g$, $f \vee g$, $f \wedge g$, and af also belong to $\mathfrak{F}_+^\#$.

Because for every function f on X the system $\{g \in \mathfrak{F} : g \leq f\}$ is either empty or directed upwards, we also have

(2.16) $$f \in \mathfrak{F}_+^\# \Longleftrightarrow f = \sup \{g : g \in \mathfrak{F},\ g \leq f\} > -\infty$$

(2.17) $f \in \mathfrak{F}_+^\#$ iff there is an arbitrary (i.e., not necessarily directed) nonempty system $\Phi \subset \mathfrak{F}$ such that $f = \sup \{g : g \in \Phi\}$.

The next proposition is an analogue of Proposition (2.10). It shows that $\mathfrak{F}_+^\#$ is closed with respect to taking limits of upwards-directed systems.

(2.18) Proposition. Let $\Phi \subset \mathfrak{F}_+^\#$ and $\Phi \nearrow f$. If $\Psi = \bigcup_{g \in \Phi} \{h \in \mathfrak{F} : h \leq g\}$, then $\Psi \nearrow f$. In particular, $f \in \mathfrak{F}_+^\#$.

Proof. Choose $h_1, h_2 \in \Psi$. Then $h_i \leq g_i$ for some $g_i \in \Phi$, $i = 1, 2$. Because Φ is directed upwards, there is a $g \in \Phi$ such that $g_1 \vee g_2 \leq g$. Hence $h_1 \vee h_2 \leq g$ also, and by (2.2) we have $h_1 \vee h_2 \in \Psi$. Since Ψ is nonempty, it follows that it is directed upwards. Moreover, by (2.16)

$$\sup \{h : h \in \Psi\} = \sup \{\sup \{h : h \in \mathfrak{F},\ h \leq g\} : g \in \Phi\}$$
$$= \sup \{g : g \in \Phi\} = f \quad ///$$

(2.19) Corollary. If $\emptyset \neq \Phi \subset \mathfrak{F}_+^\#$ and $f = \sup\{g : g \in \Phi\}$, then $f \in \mathfrak{F}_+^\#$.

Statements analogous to (2.7)-(2.11) and (2.15)-(2.19) hold also for $\mathfrak{F}_-^*$ and $\mathfrak{F}_-^\#$, respectively. The formulation of these statements is an easy kind of dualization. They can be proved either by dualizing proofs of the original statements or by using (2.5) and (2.13). Further on we shall encounter this type of duality in many places. Without explicitly mentioning it, we shall always leave the formulation and proofs of dual statements to the reader.

We shall close this chapter by giving names to the extensions we have defined.

(2.20) Definition. Let X be a set and $\mathfrak{F}$ a fundamental system on X. The systems

$$\bar{\mathfrak{F}}^* = \mathfrak{F}_+^* \cup \mathfrak{F}_-^* \quad \text{and} \quad \bar{\mathfrak{F}}^\# = \mathfrak{F}_+^\# \cup \mathfrak{F}_-^\#$$

are called the Daniell and the Bourbaki extensions of $\mathfrak{F}$, respectively.

The relationship between the Daniell and Bourbaki extensions of $\mathfrak{F}$ is simple:

(2.21) $$\mathfrak{F}_+^* \subset \mathfrak{F}_+^\# \quad \text{and} \quad \mathfrak{F}_-^* \subset \mathfrak{F}_-^\#$$

Indeed, every monotone sequence is also a directed system with the appropriate direction: upwards for an increasing sequence and downwards for a decreasing sequence.

Exercises

In exercises (2-1) through (2-7), X denotes an arbitrary set.

(2-1) Show that the system $\mathfrak{F}$ of all bounded functions on X is a fundamental system on X, and prove that $\bar{\mathfrak{F}}^* = \bar{\mathfrak{F}}^\#$ and $\mathfrak{F} = \mathfrak{F}_+^* \cap \mathfrak{F}_-^*$.

(2-2)* Let $\mathfrak{F}$ consist of all real-valued functions f on X for which the set f(X) is finite. Show that $\mathfrak{F}$ is a fundamental system on X, and prove

(i) $\mathfrak{F}_+^*$ consists of all functions on X which are bounded from below. Hint: Assume $f \geq 0$, choose an integer $n \geq 1$, and set $A_0 = \{x \in X : f(x) \geq n\}$ and $A_i = \{x \in X : (i-1)2^{-n} \leq f(x) < i2^{-n}\}$, $i = 1, \ldots, n2^n$. Let $f_n(x) = n$ for $x \in A_0$, and $f_n(x) = (i-1)2^{-n}$ for $x \in A_i$, $i = 1, \ldots, n2^n$. Draw a picture and show that $f_n \nearrow f$.

(ii) $\bar{\mathfrak{F}}^* = \bar{\mathfrak{F}}^\#$.

(2-3)* Let $\aleph$ be an infinite cardinal, and let $\mathfrak{F}$ consist of all real-valued functions f on X for which $\|f(X)\| \leq \aleph$. Show that $\mathfrak{F}$ is a fundamental system on X, and prove

(i) $\mathfrak{F}_+^*$ consists of all functions f on X for which $f > -\infty$. Hint: Modify the technique hinted in (2-2)(i).

(ii) $\bar{\mathfrak{F}}^* = \bar{\mathfrak{F}}^\#$.

(iii) $\mathfrak{F} = \mathfrak{F}_+^* \cap \mathfrak{F}_-^*$ iff $\aleph \geq c$ or $\|X\| \leq \aleph$.

(2-4) Let $\mathfrak{F}$ consist of all real-valued functions f on X for which there is a $c_f \in \mathbf{R}$ such that the set $\{x \in X : f(x) \neq c_f\}$ is finite. Show that $\mathfrak{F}$ is a fundamental system on X, and prove

(i) $\mathfrak{F} = \mathfrak{F}_+^* \cap \mathfrak{F}_-^*$ iff X is finite.

(ii) $\bar{\mathfrak{F}}^* = \bar{\mathfrak{F}}^\#$ iff X is countable.

(2-5) Let $\aleph$ be an infinite cardinal, and let $\mathfrak{F}$ consist of all real-valued functions on X for which there is a $c_f \in \mathbf{R}$ such that

$$\|\{x \in X : f(x) \neq c_f\}\| \leq \aleph$$

Show that $\mathfrak{F}$ is a fundamental system on X, and prove

(i) $\mathfrak{F} = \mathfrak{F}_+^* \cap \mathfrak{F}_-^*$.

(ii) $\bar{\mathfrak{F}}^* = \bar{\mathfrak{F}}^\#$ iff $\|X\| \leq \aleph$.

(iii) $\mathfrak{F} = \mathfrak{F}_+^\# \cap \mathfrak{F}_-^\#$ iff $\|X\| \leq \aleph$.

(2-6)$^+$ Let $\mathfrak{F}$ consist of all real-valued functions f on X for which the set $\{x \in X : f(x) \neq 0\}$ is finite. Show that $\mathfrak{F}$ is a fundamental system on X, and prove

(i) $\mathfrak{F} = \mathfrak{F}_+^* \cap \mathfrak{F}_-^* = \mathfrak{F}_+^\# \cap \mathfrak{F}_-^\#$.

(ii) $\bar{\mathfrak{F}}^* = \bar{\mathfrak{F}}^\#$ iff X is countable.

(2-7) Let $\mathfrak{F}$ consist of all real-valued functions f on X for which

$$\| \{x \in X : f(x) \neq 0\} \| < \|X\|$$

Show that $\mathfrak{F}$ is a fundamental system on X iff X is infinite. Assuming X is infinite, prove

(i) $\mathfrak{F} = \mathfrak{F}_+^\# \cap \mathfrak{F}_-^\#$.

(ii) $\bar{\mathfrak{F}}^* = \bar{\mathfrak{F}}^\#$ iff $\|X\| = \aleph_\alpha$, where α is cofinal with ω [see (1-11)].

(2-8)$^+$ Let X = [0,1]. Show that the following systems of functions on X are fundamental systems on X.

(a) $\mathfrak{F}_1$ consists of all continuous functions f on X for which $f(0) = 0$ and $f'_+(0)$ exists.

(b) $\mathfrak{F}_2$ consists of all linear functions f on X for which $f(0) = 0$.

Prove

(i) $\bar{\mathfrak{F}}_2^* = \bar{\mathfrak{F}}_2^\#$.

(ii) $f \in \mathfrak{F}_{2+}^* - \mathfrak{F}_2$ iff $f(0) = 0$ and $f(x) = +\infty$ for $x \in (0,1]$.

(2-9) Let X = **R**. Which of the following systems of functions on X are fundamental systems on X?

(i) $\mathfrak{F}_1$ consists of all finite continuous functions.

(ii) $\mathfrak{F}_2$ consists of all Lipschitz functions [a function f on **R** is called Lipschitz if there is an $\alpha \in \mathbf{R}$ such that $|f(x) - f(y)| \leq \alpha |x - y|$ for all $x, y \in \mathbf{R}$].

(iii) $\mathfrak{F}_3$ consists of all differentiable functions.

(iv) $\mathfrak{F}_4$ consists of all polynomials.

(v) $\mathfrak{F}_5$ consists of all finite continuous functions of polynomial growth [a function f on $\mathbf{R}$ is said to be of polynomial growth whenever $\lim_{x\to\pm\infty} x^{-n}f(x) = 0$ for some positive integer n].

(vi) $\mathfrak{F}_6$ consists of all finite continuous functions vanishing at infinity [a function f on $\mathbf{R}$ is said to vanish at infinity whenever $\lim_{x\to\pm\infty} f(x) = 0$].

(vii) $\mathfrak{F}_7$ consists of all finite continuous functions vanishing exponentially at infinity [a function is said to vanish exponentially at infinity whenever $\lim_{x\to\pm\infty} e^{\alpha|x|}f(x) = 0$ for some positive $\alpha \in \mathbf{R}$].

(viii) $\mathfrak{F}_8$ consists of all finite continuous functions which are zero outside of a compact interval.

(2-10)*+ Let $a,b \in \mathbf{R}$, $a < b$, and let $X = [a,b]$. Show that the following systems of functions on X are fundamental systems on X.

(a) $\mathfrak{F}_1$ is the system of all finite continuous functions.

(b) $\mathfrak{F}_2$ is the system of all piecewise linear functions (a finite continuous function on $[a,b]$ is called piecewise linear if its graph is a finite polygon).

(c) $\mathfrak{F}_3$ is the system of all elementary functions (a function f on $[a,b]$ is called an elementary function if there are disjoint, possibly degenerate, intervals $J_1, \ldots, J_n$ and real numbers $c_1, \ldots, c_n$ such that $\bigcup_{i=1}^n J_i = [a,b]$ and $f(x) = c_i$ for all $x \in J_i$, $i = 1, \ldots, n$).

Prove

(i) $\mathfrak{F}_1^* = \mathfrak{F}_2^*$ and $\mathfrak{F}_1^\# = \mathfrak{F}_2^\#$.

(ii) $\bar{\mathfrak{F}}_1^* \subset \bar{\mathfrak{F}}_3^*$ and $\bar{\mathfrak{F}}_1^\# \subset \bar{\mathfrak{F}}_3^\#$.

(iii) $\mathfrak{F}_3 \neq \mathfrak{F}_{3+}^* \cap \mathfrak{F}_{3-}^*$.

(iv) $\bar{\mathfrak{F}}_3^* \neq \bar{\mathfrak{F}}_3^\#$. Hint: For $x \in [a,b]$ let $f(x) = 0$ if x is rational, and $f(x) = 1$ if x is irrational, and show that $f \in \mathfrak{F}_{3+}^\# - \mathfrak{F}_{3+}^*$.

(2-11)$^+$ Let $a, b \in \mathbf{R}$, $a < b$, and let $X = [a, b)$. A function f on X is called a step function if there are real numbers $x_0, x_1, \ldots, x_n$ and $c_1, \ldots, c_n$ such that $a = x_0 < x_1 < \cdots < x_n = b$ and $f(x) = c_i$ for all $x \in [x_{i-1}, x_i)$, $i = 1, \ldots, n$. Clearly, every step function on X is also an elementary function on X [see (2-10)(c)], but not vice versa. Let $\mathfrak{F}$ consist of all step functions on X. Show that $\mathfrak{F}$ is a fundamental system on X and that $\mathfrak{F} \neq \mathfrak{F}_+^* \cap \mathfrak{F}_-^*$.

3. THE FUNDAMENTAL SYSTEM OF FUNCTIONS IN A TOPOLOGICAL SPACE

We shall study the important case when there is a topology defined on X. Although, larger generality is possible we shall restrict ourselves to locally compact Hausdorff spaces only.

Let X be a topological space. If f is a function on X, we let

$$\operatorname{supp} f = \{x \in X : f(x) \neq 0\}^{-}$$

and we call this set the support of f. By $\mathcal{C}(X)$ we denote the family of all continuous real-valued functions on X, and by $\mathcal{C}_0(X)$ we denote the family of all $f \in \mathcal{C}(X)$ for which supp f is compact. Clearly, both $\mathcal{C}(X)$ and $\mathcal{C}_0(X)$ satisfy conditions (i)-(iii) of Definition (2.1) and thus both can serve as fundamental systems on X. It turns out, however, that for locally compact X the one of crucial importance is $\mathcal{C}_0(X)$. A reason for this comes up when $X = \mathbf{R}$: There is no natural way to extend the Riemann integral to the class of all finite continuous functions on $\mathbf{R}$.

Throughout this chapter we shall assume that X is a locally compact Hausdorff space and that $\mathfrak{F} = \mathcal{C}_0(X)$ is the fundamental system on X. Our main task will be to describe the Daniell and Bourbaki extensions of $\mathfrak{F}$ and, in terms of the topology on X, to give a necessary and sufficient condition under which they coincide.

If $f \in \mathcal{C}_0(X)$ and $g \in \mathcal{C}(X)$, then clearly $fg \in \mathcal{C}_0(X)$. The following observations are easy consequences of this fact.

(3.1) Let g be a nonnegative function from $\mathcal{C}(X)$ or $\mathfrak{F}_+^*$. Then $fg \in \mathfrak{F}_+^*$ whenever $f \in \mathfrak{F}_+^*$.

(3.2) Let g be a nonnegative function from $\mathcal{C}(X)$ or $\mathfrak{F}_+^{\#}$. Then $fg \in \mathfrak{F}_+^{\#}$ whenever $f \in \mathfrak{F}_+^{\#}$.

(3.3) Definition. A function f on X is said to be lower semicontinuous at $x \in X$, if for every $\alpha < f(x)$ there is a neighborhood U of x such that

$$y \in U \Rightarrow f(y) > \alpha$$

A function f on X which is lower semicontinuous at every $x \in X$ is called lower semicontinuous on X.

The reader should formulate the dual definition of upper semicontinuity and show that a function on X is continuous iff it is both upper and lower semicontinuous.

(3.4) Lemma. Let f be a lower semicontinuous function on X, and let $F \subset X$ be a compact set. If there is an $\alpha \in \mathbf{R}^-$ such that $f(x) > \alpha$ for all $x \in F$, then also $\inf\{f(x) : x \in F\} > \alpha$.

Proof. For every $x \in F$ there is an α_x, $\alpha < \alpha_x < f(x)$, and an open neighborhood U_x of x such that

$$y \in U_x \Rightarrow f(y) > \alpha_x$$

Since $\{U_x : x \in F\}$ is an open cover of the compact set F, we can find $x_1, \ldots, x_n$ in F such that $F \subset \bigcup_{i=1}^n U_{x_i}$. Thus, if $\alpha_0 = \wedge_{i=1}^n \alpha_{x_i}$, we have $f(x) > \alpha_0 > \alpha$ for all $x \in F$, and the lemma follows. ///

(3.5) Proposition. A function f on X belongs to $\mathfrak{F}_+^\#$ iff

(i) $f > -\infty$

(ii) The set $\{x \in X : f(x) < 0\}^-$ is compact

(iii) f is lower semicontinuous.

Proof. Let $f \in \mathfrak{F}_+^\#$. Then by (2.16)

$$f = \sup\{g : g \in \mathfrak{F},\ g \leq f\} > -\infty$$

Thus there is a $g_0 \in \mathfrak{F}$ for which $g_0 \leq f$. This implies (ii) since $\{x \in X : f(x) < 0\}^-$ is contained in supp g_0 which is compact. To prove (iii), choose $x \in X$ and $\alpha < f(x)$. There is a $g_1 \in \mathfrak{F}$ such that $g_1 \leq f$ and $\alpha < g_1(x)$. Since g_1 is continuous we can find a neighborhood U of x such that

$$y \in U \Rightarrow f(y) \geq g_1(y) > \alpha$$

To prove the converse, let f be a function on X satisfying (i)-(iii) and let

$$g = \sup\{h : h \in \mathfrak{F},\ h \leq f\}$$

Clearly, $g \leq f$, and so according to (2.16), it suffices to show that $g \geq f$. We shall do this by applying twice the Urysohn lemma (see Ref. 16, Chapter 4, Lemma 4, p. 115 together with Chapter 5, Theorem 9, p. 141). The idea of the proof is illustrated in Fig. 3.1. According to (3.4), $f > \alpha$ for some $\alpha \in \mathbf{R}$. Choose $x_0 \in X$ and $\beta \in \mathbf{R}$, $\alpha < \beta < f(x_0)$. By (iii) there is an open neighborhood U of x_0 such that U^- is compact and

$$x \in U \Rightarrow f(x) > \beta$$

Using the Urysohn lemma, we can find a function $\phi \in C(X)$ such that $a \leq \phi \leq \beta$, $\phi(x_0) = \beta$ and $\phi(x) = \alpha$ for all $x \in X - U$. Although $\phi \leq f$, the function ϕ has to be modified since its support may not be compact. By (ii) the set $F = U^- \cup \{x \in X : f(x) < 0\}^-$ is compact. Since X is Hausdorff and locally compact, there is an open set V containing F for which V^- is compact. Using the Urysohn lemma again, we can find a function $\psi \in C(X)$ such that $0 \leq \psi \leq 1$, $\psi(x) = 1$ for all $x \in F$, and $\psi(x) = 0$ for all $x \in X - V$. Clearly, ψ belongs to $\mathfrak{F}$ and so does $h = \phi\psi$. It is easy to see that $h \leq f$. Since $h(x_0) = \phi(x_0) = \beta$, $g(x_0) \geq \beta$. It follows from the arbitrariness of β that $g(x_0) \geq f(x_0)$. ///

(3.6) Corollary. $\mathfrak{F} = \mathfrak{F}_+^* \cap \mathfrak{F}_-^* = \mathfrak{F}_+^\# \cap \mathfrak{F}_-^\#$.

This corollary follows from (2.21), the previous proposition, and its dual formulation for $\mathfrak{F}_-^\#$. As an immediate consequence we obtain the following corollary.

(3.7) Corollary. $\mathfrak{F}^\# = \mathfrak{F}^*$ iff $\mathfrak{F}_+^\# = \mathfrak{F}_+^*$ and $\mathfrak{F}_-^\# = \mathfrak{F}_-^*$.

The structure of lower semicontinuous functions is, in general, quite complicated. Thus it is useful to introduce a family of simpler functions by which lower semicontinuous functions can be approx-

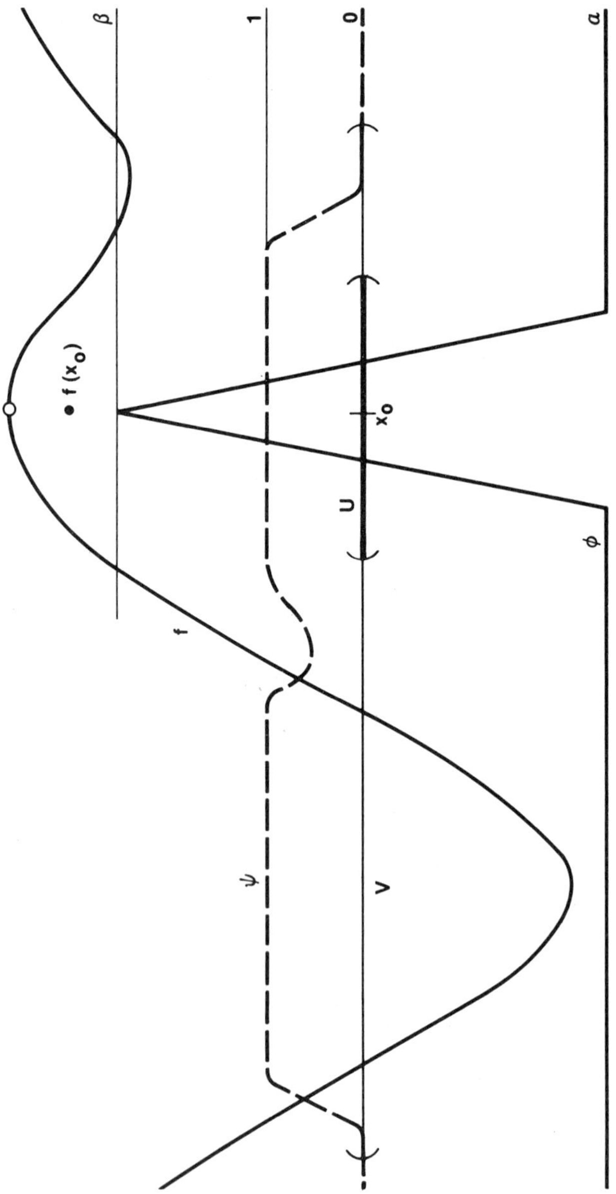

FIG. 3.1.

imated. This kind of approach is pretty standard; we shall use it again in various contexts.

(3.8) Definition. A function f on X is called steplike whenever there are open sets $\emptyset = G_0 \subset G_1 \subset \cdots \subset G_n = X$ and real numbers $a_1 > a_2 > \cdots > a_n$ such that $f(x) = a_i$ for all $x \in G_i - G_{i-1}$, $i = 1, \ldots, n$.

It is easy to see that every steplike function is lower semicontinuous (see Fig. 3.2).

(3.9) Lemma. Let f be a bounded, lower semicontinuous function on X. Then there is a sequence $\{f_n\}$ of steplike functions such that $f_n \nearrow f$.

Proof. Since f is bounded there are $a, b \in \mathbf{R}$ such that $a < f < b$. Choose an integer $n \geq 1$, and let $k_{n,i} = b - i2^{-n}(b - a)$ and $G_{n,i} = \{x \in X : f(x) > k_{n,i}\}$, $i = 0, 1, \ldots, 2^n$. Clearly, $\emptyset = G_{n,0} \subset G_{n,1} \subset \cdots \subset G_{n,2^n} = X$, and it follows from the semicontinuity of f that the sets $G_{n,i}$ are open. Thus if for $x \in G_{n,i} - G_{n,i-1}$ we set $f_n(x) = k_{n,i}$, $i = 1, \ldots, 2^n$, we see that f_n is a steplike function on X. Moreover, $f_n \leq f_{n+1} \leq f$ and $f - f_n \leq 2^{-n}(b - a)$. ///

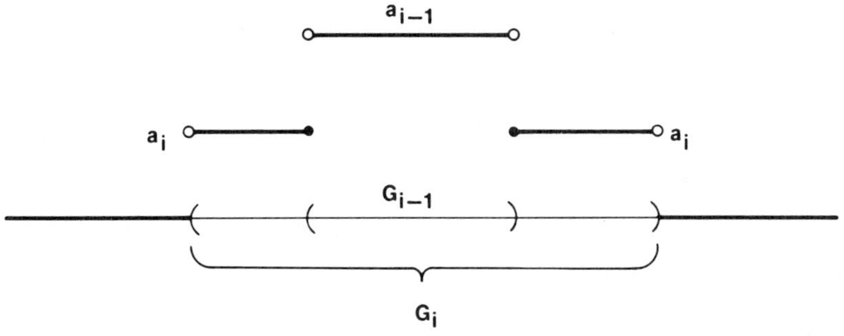

FIG. 3.2.

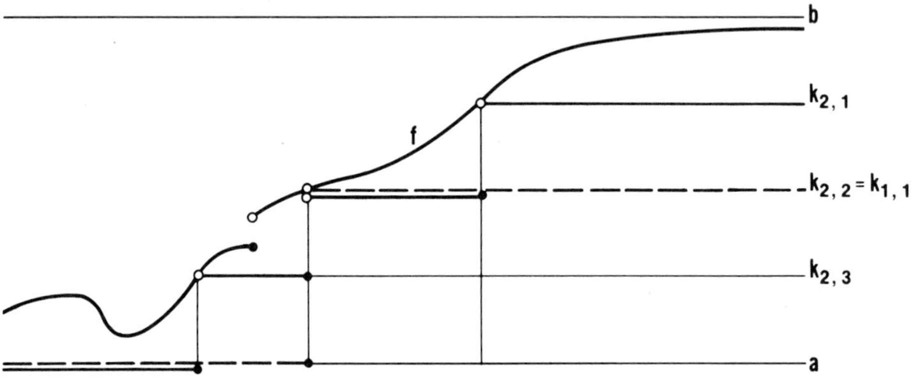

FIG. 3.3.

The idea of the previous proof is illustrated in Fig. 3.3.

(3.10) Lemma. Let G be an open, σ-compact subset of X, and let $a, b \in \mathbf{R}$, $a < b$. If $f(x) = b$ for $x \in G$ and $f(x) = a$ for $x \in X - G$, then there is a sequence $\{f_n\} \subset \mathcal{C}(X)$ such that $f_n \nearrow f$.

Proof. Since G is σ-compact, there is a sequence $\{F_n\}$ of compact sets such that $G = \bigcup_{n=1}^{\infty} F_n$. Thus we can find functions $\varphi_n \in \mathcal{C}(X)$ such that $a \leq \varphi_n \leq b$, $\varphi_n(x) = b$ if $x \in F_n$, and $\varphi_n(x) = a$ if $x \in X - G$, $n = 1, 2, \ldots.$ The proof is completed by letting $f_n = \vee_{i=1}^{n} \varphi_i$, $n = 1, 2, \ldots.$ ///

(3.11) Corollary. Let f be a steplike function on X, and let every open subset of X be σ-compact. Then there is a sequence $\{f_n\} \subset \mathcal{C}(X)$ such that $f_n \nearrow f$.

Proof. Let $\emptyset = G_0 \subset G_1 \subset \cdots \subset G_k = X$ be open sets and $a_1 > a_2 > \cdots > a_k$ be real numbers such that $f(x) = a_i$ for $x \in G_i - G_{i-1}$, $i = 1, \ldots, k$. Setting $f_i(x) = a_i$ if $x \in G_i$ and $f_i(x) = a_k$ if $x \in X - G_i$, we have $f = \vee_{i=1}^{k-1} f_i$ (see Fig. 3.2). By (3.10) there are sequences $\{f_{i,n}\}_{n=1}^{\infty} \subset \mathcal{C}(X)$ such that $f_{i,n} \nearrow f_i$ as $n \to +\infty$, $i = 1, \ldots, k - 1$. Thus it suffices to let $f_n = \vee_{i=1}^{k-1} f_{i,n}$, $n = 1, 2, \ldots.$ ///

(3.12) Corollary. Let f be a bounded lower semicontinuous function on X, and let every open subset of X be σ-compact. Then there is a sequence $\{f_n\} \subset \mathcal{C}(X)$ such that $f_n \nearrow f$.

This corollary follows from (3.9) and (3.11) by the same diagonal argument that was used to prove Proposition (2.10). We shall leave it to the reader as an exercise to carry the proof out formally.

(3.13) Theorem. The Daniell and Bourbaki extensions of $\mathfrak{F} = \mathcal{C}_o(X)$ coincide iff every open subset of X is σ-compact.

Proof. Let $\mathfrak{F}^* = \mathfrak{F}^\#$, and let $G \subset X$ be open. For $x \in X$, set $f(x) = 1$ if $x \in G$ and $f(x) = 0$ if $x \in X - G$. Clearly, f is a nonnegative lower semicontinuous function on X. By (3.5) $f \in \mathfrak{F}_+^\#$, and by (3.7) there is a sequence $\{f_n\} \subset \mathfrak{F}$ such that $f_n \nearrow f$. The sets

$$F_n = \{x \in \text{supp}\, f_n : f_n(x) \geq 1/2\}$$

are compact and $G = \bigcup_{n=1}^{\infty} F_n$.

Conversely, suppose that every open subset of X is σ-compact, and choose $f \in \mathfrak{F}_+^\#$. It follows from (3.5) and (3.4) that $f \geq a$ for some $a \in \mathbf{R}$. Since X is locally compact and σ-compact, there are open sets $U_n \subset X$ such that U_n^- is compact, $U_n^- \subset U_{n+1}$, $n = 1, 2, \ldots$, and $X = \bigcup_{n=1}^{\infty} U_n$. Thus we can find functions $\varphi_n \in \mathcal{C}(X)$, $n = 1, 2, \ldots$, such that $0 \leq \varphi_n \leq n$, $\varphi_n(x) = n$ if $x \in U_n^-$, and $\varphi_n(x) = 0$ if $x \in X - U_{n+1}$. Since $\varphi_n \in \mathfrak{F}$, it follows from (2.15) that

$$f_n = f \wedge (\vee_{i=1}^{n} \varphi_n)$$

belongs to $\mathfrak{F}_+^\#$. Clearly, $a^- \leq f_n \leq n$ and $f_n \nearrow f$. By (3.12) there are sequences $\{g_{n,k}\}_{k=1}^{\infty} \subset \mathcal{C}(X)$, $n = 1, 2, \ldots$, for which $g_{n,k} \nearrow f_n$ as $k \to +\infty$. Moreover,

$$\text{supp}\, f_n \subset U_{n+1}^- \cup \{x \in X : f(x) < 0\}^-$$

and so by (3.5)(ii), supp f_n is compact. Choose an open set $V_n \subset X$ for which V_n^- is compact and supp $f_n \subset V_n$. There are functions

$\psi_n \in \mathcal{C}(X)$, $n = 1, 2, \ldots$, such that $0 \le \psi_n \le 1$, $\psi_n(x) = 1$ if $x \in \operatorname{supp} f_n$, and $\psi_n(x) = 0$ if $x \in X - V_n$. Letting $h_{n,k} = \psi_n g_{n,k}$, we have $\{h_{n,k}\}_{k=1}^{\infty} \subset \mathfrak{F}$, $n = 1, 2, \ldots$, and $h_{n,k} \nearrow f_n$ as $k \to +\infty$. Therefore $f_n \in \mathfrak{F}_+^*$, and it follows from (2.10) that $f \in \mathfrak{F}_+^*$ also. Hence $\mathfrak{F}_+^{\#} = \mathfrak{F}_+^*$ by (2.21). Using (2.5) and (2.13), we also obtain $\mathfrak{F}_-^{\#} = \mathfrak{F}_-^*$. ///

We shall close this chapter with an important lemma which we shall use later.

(3.14) Lemma. Let $\Phi \subset \mathfrak{F}$, and let $\Phi \searrow 0$. Then for every $\varepsilon > 0$, there is an $f_\varepsilon \in \Phi$ such that $f_\varepsilon < \varepsilon$.

Proof. Choose $\varepsilon > 0$ and $f_0 \in \Phi$. If $f \in \Phi$, let $U_f = \{x \in X : f(x) < \varepsilon\}$. Because $\Phi \searrow 0$, $\{U_f : f \in \Phi\}$ is an open cover of X. Since $\operatorname{supp} f_0$ is compact, there are $f_1, \ldots, f_n$ in Φ such that

$$\operatorname{supp} f_0 \subset \bigcup_{i=1}^{n} U_{f_i}$$

It suffices to choose $f_\varepsilon \in \Phi$ for which $f_\varepsilon \le \bigwedge_{i=0}^{n} f_i$. ///

Exercises

In exercises (3-1) through (3-4) we shall assume that $\mathfrak{F} = \mathcal{C}_0(X)$.

(3-1)$^+$ Let X be a discrete space. Show that $\mathfrak{F}$ consists of all real-valued functions f on X for which the set $\{x \in X : f(x) \neq 0\}$ is finite [see (2-6)].

(3-2) Let $X = W^-$ [see Chapter 1, Section B], and let $f(x) = 1$ if $x \in W$ and $f(\Omega) = 0$. Show that $f \in \mathfrak{F}_+^{\#}$ and $f \notin \mathfrak{F}_+^*$. Hint: Use (1-15)(ii).

(3-3)* Let $I = [0,1]$, and let $X = I \times \{0,1\}$. Show that in X we can define a topology as follows: Points (s,1) are isolated and a neighborhood base at (s,0) is given by sets

$$(U \times \{0\}) \cup [(U - \{x\}) \times \{1\}]$$

where U is a neighborhood of s in the ordinary topology of I (draw a picture). Give X this topology, and prove

(i) X is a first countable, compact Hausdorff space.

(ii) $I \times \{1\}$ is an open subset of X which is not σ-compact.

(iii) If $f(s,t) = t$ for each $(s,t) \in X$, then $f \in \mathfrak{F}_+^{\#}$ and $f \notin \mathfrak{F}_+^*$.

(3-4)$^+$ Dini theorem. Let X be a locally compact Hausdorff space. Prove that if a monotone sequence $\{f_n\} \subset \mathfrak{F}$ converges to $f \in \mathfrak{F}$, then it converges uniformly. Hint: Use (3.14).

(3-5)* Show that a normal space X is perfectly normal [see (1-12)] iff for each bounded, lower semicontinuous function f on X, there is a sequence $\{f_n\} \subset C(X)$ such that $f_n \nearrow f$. Hint: Modify the proof of (3.12) and the first part of the proof of (3.13).

(3-6)$^+$ Let X be a metrizable space. Show that every open subset of X is σ-compact iff X is σ-compact. Hint: See (1-12).

(3-7)* Let X be a metrizable space, and let f be a bounded, lower semicontinuous function on X. Prove directly that there is a sequence $\{f_n\} \subset C(X)$ such that $f_n \nearrow f$. Hint: Choose a metric ρ on X and set

$$f_n(x) = \inf\{f(y) + n\rho(x,y) : y \in X\}$$

for $x \in X$ and $n = 1, 2, \ldots$.

(3-8)* Let X be a locally compact metric space with metric ρ. A function f on X is called Lipschitz if there is a constant $\alpha \in \mathbf{R}$ such that $|f(x) - f(y)| \leq \alpha\rho(x,y)$ for all $x,y \in X$. Let $\mathfrak{F}_1 = C_o(X)$, and let $\mathfrak{F}_2$ consist of all Lipschitz functions from $\mathfrak{F}_1$. Show that

(i) $\mathfrak{F}_2 = \mathfrak{F}_1$ iff X is discrete. Hint: If X is not discrete, choose $\{x_n\} \subset X$ converging to $x_\infty \in X$ with the property that $\alpha_n = \rho(x_n, x_{n+1}) > 0$, $n = 1, 2, \ldots$, and $\Sigma_{n=1}^{\infty} \sqrt{\alpha_n} < +\infty$. For $n = 1, 2, \ldots, \infty$, set

$f(x_n) = \Sigma_{k=1}^{n} \sqrt{\alpha_k}$, and use the Tietze extension theorem (see Ref. 16, Chapter 7, problem O, p. 242) and the Urysohn lemma to extend f to a function $\hat{f} \in C_o(X)$. Observe that $\hat{f} \notin \mathfrak{F}_2$.

(ii) $\bar{\mathfrak{F}}_2^* = \bar{\mathfrak{F}}_1^*$. Hint: Observe that the functions f_n hinted in (3-7) are Lipschitz.

(iii) $\bar{\mathfrak{F}}_2^* = \bar{\mathfrak{F}}_2^{\#}$ iff X is σ-compact. Hint: Use (ii), (3-6), and (3.13).

(3-9) Let X = **R**, and consider the fundamental systems $\mathfrak{F}_i$, i = 1, 2, 5, 6, 7, 8, from (2-9). Show that

(i) $\bar{\mathfrak{F}}_i^* = \bar{\mathfrak{F}}_i^{\#}$, i = 1, 2, 5, 6, 7, 8.

(ii) $\bar{\mathfrak{F}}_8^* \subsetneqq \bar{\mathfrak{F}}_7^* \subsetneqq \bar{\mathfrak{F}}_6^* \subsetneqq \bar{\mathfrak{F}}_5^* \subsetneqq \bar{\mathfrak{F}}_1^*$.

(iii) $\bar{\mathfrak{F}}_6^* \subsetneqq \bar{\mathfrak{F}}_2^* \subsetneqq \bar{\mathfrak{F}}_5^*$.

(3-10) Let a,b ∈ **R**, a < b, and let X = [a,b]. Suppose that $\mathfrak{F}$ consists of all elementary functions on X [see (2-10)(c)]. Prove

(i) $\mathfrak{F}_+^*$ consists of all functions f on X which are bounded from below and are lower semicontinuous at all but countably many x ∈ X.

(ii) $\mathfrak{F}_+^{\#}$ consists of all functions f on X which are bounded from below.

(3-11)* Let a,b ∈ **R**, a < b, and let X = [a,b). Suppose that $\mathfrak{F}$ consists of all step functions on X [see (2-11)]. A function f on X is said to be lower semicontinuous from the right at x ∈ X if for every $\alpha < f(x)$, there is a $\delta > 0$ such that $\alpha < f(y)$ for all $y \in X \cap [x, x + \delta)$. Lower semicontinuity from the left is defined analogously. Prove

(i) $\mathfrak{F}_+^{\#}$ consists of all functions on X which are bounded from below and are lower semicontinuous from the right at all x ∈ X.

(ii) $\bar{\mathfrak{F}}^* = \bar{\mathfrak{F}}^{\#}$. Hint: Give X the half-open interval topology [see Ref. 16, Chapter 1, problem K, p. 59]. In this topology each function from $\mathfrak{F}_+^{\#}$ is lower semicontinuous. Modify the proof of (3.9) to show that for each $f \in \mathfrak{F}_+^{\#}$, there is a sequence $\{f_n\}$ of steplike functions [see (3.8)] such that $f_n \nearrow f$. Using the fact that the half-open interval topology is hereditarily Lindelof, observe that all steplike functions belong to $\mathfrak{F}_+^*$.

(iii) If a function f on X is lower semicontinuous from the right at all $x \in X$, then it is lower semicontinuous from the left at all but countably many $x \in X$. Hint: Apply (i) and (ii) to functions $f \vee (-n)$, $n = 1, 2, \ldots$.

Note: A direct proof of (iii) is possible by using (5-8).

(3-12)* Let X be an arbitrary Lindelof space, $\Phi \subset \mathcal{C}(X)$, and let $\Phi \searrow 0$. Show that there is a sequence $\{f_n\} \subset \Phi$ such that $f_n \searrow 0$. Hint: Modify the proof of (3.14).

(3-13)+ Baire functions. Let X be any topological space, and let $\mathcal{B}_{o0} = \mathcal{C}(X)$. Let $\alpha < \Omega$ be an ordinal, and suppose that we have already defined families $\mathcal{B}_{o\beta}$ for all ordinals $\beta < \alpha$. Then $\mathcal{B}_{o\alpha}$ consists of all functions f on X for which there is a sequence $\{f_n\} \subset \bigcup_{\beta<\alpha} \mathcal{B}_{o\beta}$ such that $f_n \to f$. Functions from the family $\mathcal{B}_o = \bigcup_{\alpha<\Omega} \mathcal{B}_{o\alpha}$ are called Baire functions on X. Show that

(i) Finite functions from $\mathcal{B}_o$ form a fundamental system on X.

(ii) $\mathcal{B}_o$ is closed with respect to taking limits, i.e., if $\{f_n\} \subset \mathcal{B}_o$ and $f_n \to f$ for some function f on X, then $f \in \mathcal{B}_o$.

(iii) $\mathcal{B}_o$ is the smallest family of functions on X which is closed with respect to taking limits and contains $\mathcal{C}(X)$; more precisely, if $\mathcal{B}$ is a family of functions on X which is closed with respect to taking limits and if $\mathcal{C}(X) \subset \mathcal{B}$, then also $\mathcal{B}_o \subset \mathcal{B}$.

Note: In general, it may happen that $\mathcal{B}_o = \mathcal{B}_{o\alpha}$ for some ordinal $\alpha < \Omega$. For instance, if X is a discrete space, then $\mathcal{B}_o = \mathcal{B}_{o0}$. Thus it is a remarkable result of Lebesgue that if $X = \mathbf{R}$, then $\mathcal{B}_{o\alpha} \subsetneqq \mathcal{B}_{o\beta}$ for each $\alpha < \beta < \Omega$ (see, e.g., Ref. 19, Chapter 15, Section 2).

(3-14)+ Let f and g be finite Baire functions on a topological space X, and let h be a Baire function on $\mathbf{R}^2$. For each $x \in X$, set

$$H(x) = h[f(x), g(x)]$$

(i) Show that H is a Baire function on X. Hint: Show it for

$h \in C(R^2)$ and then proceed by transfinite induction.

(ii) Conclude from (i) that the system $\mathfrak{F}$ of all finite Baire functions on X is a fundamental system on X.

(iii) Use (i) to show that $fg \in \mathfrak{F}$ whenever $f \in \mathfrak{F}$ and $g \in \mathfrak{F}$.

4. INTEGRALS ON THE FUNDAMENTAL SYSTEM AND THEIR FIRST EXTENSIONS

In Chapter 2, when we introduced the notion of a fundamental system $\mathfrak{F}$, we mentioned that $\mathfrak{F}$ is a basic system of functions on which an integral can be defined. At this point, however, it is not clear how to define an integral. It is also not clear how many different integrals we can define on a given fundamental system. It turns out that, in fact, we usually have a large variety of integrals to choose from [the reader familiar with the Riemann-Stieltjes integral (see Ref. 1, Chapter 9) will have no problem seeing this]. Again, our aim will be to describe the situation in complete generality. Thus we shall define the integral quite abstractly as a functional on $\mathfrak{F}$ which satisfies certain properties. The example of the Riemann integral $\int_a^b f(x)\,dx$ of a continuous function f over a compact interval [a,b] should be the main guide to the reader's intuition.

Unless specified otherwise, throughout this chapter X will be an arbitrary set and $\mathfrak{F}$ will be a fundamental system on X.

(4.1) Definition. A map $I : \mathfrak{F} \to \mathbf{R}$ is called a nonnegative linear functional on $\mathfrak{F}$ whenever it satisfies the following conditions:

(i) $I(f + g) = If + Ig$ for all $f,g \in \mathfrak{F}$

(ii) $I(af) = aIf$ for all $a \in \mathbf{R}$ and $f \in \mathfrak{F}$

(iii) $If \geq 0$ for all $f \in \mathfrak{F}$ for which $f \geq 0$.

Conditions (i) and (ii) imply that I is a linear functional (or a linear form) on the vector space $\mathfrak{F}$. The nonnegativity of I is expressed by condition (iii).

For a nonnegative linear functional I on $\mathfrak{F}$ we obtain

(4.2) $$f = 0 \Rightarrow If = 0$$

(4.3) $$(f, g \in \mathfrak{F},\ f \le g) \Rightarrow If \le Ig$$

Indeed, if $g \in \mathfrak{F}$, then $0 \cdot g = 0$, and by (ii),

$$I(0 \cdot g) = 0 \cdot Ig = 0$$

If $f, g \in \mathfrak{F}$ and $f \le g$, then $g - f \in \mathfrak{F}$ and $g - f \ge 0$. Thus by (iii), (i), and (ii) we have

$$0 \le I(g - f) = Ig + I(-f) = Ig - If$$

and so $If \le Ig$.

(4.4) Definition. A nonnegative linear functional I^* on $\mathfrak{F}$ is called a Daniell integral on $\mathfrak{F}$ whenever

$$(\{f_n\} \subset \mathfrak{F},\ f_n \searrow 0) \Rightarrow I^*f_n \to 0$$

(4.5) Definition. A nonnegative linear functional $I^\#$ on $\mathfrak{F}$ is called a Bourbaki integral on $\mathfrak{F}$ whenever

$$(\Phi \subset \mathfrak{F},\ \Phi \searrow 0) \Rightarrow \inf\{I^\# f : f \in \Phi\} = 0$$

Note that by (4.3) we have $I^*f_n \searrow 0$ and that $\{I^\# f : f \in \Phi\}$ is a downwards-directed set of real numbers.

For $\Phi \subset \mathfrak{F}$, we let

$$I^\# \Phi = \{I^\# f : f \in \Phi\}$$

If $\Phi \subset \mathfrak{F}$ is directed upwards and $a = \sup I^\# \Phi$, we shall write $I^\# \Phi \nearrow a$. Similarly, if $\Phi \subset \mathfrak{F}$ is directed downwards and $a = \inf I^\# \Phi$, we shall write $I^\# \Phi \searrow a$. The symbol $I^\# \Phi \to a$ means that either $I^\# \Phi \nearrow a$ or $I^\# \Phi \searrow a$. Sometimes we shall write $\lim I^\# \Phi = a$ instead of $I^\# \Phi \to a$.

As an immediate consequence of the definitions we obtain the following.

(4.6) Proposition. Every Bourbaki integral on $\mathfrak{F}$ is also a Daniell integral on $\mathfrak{F}$.

The converse of this proposition is not true. In Chapter 12 we shall introduce an important example of a Daniell integral which, in general, is not a Bourbaki integral [see also (4-2) and (4-3)].

(4.7) Theorem. Let X be a locally compact Hausdorff space, and let I be a nonnegative linear functional on $\mathfrak{F} = C_o(X)$. Then I is a Bourbaki integral on $\mathfrak{F}$.

Proof. Let $\Phi \subset \mathfrak{F}$ and $\Phi \searrow 0$. Choose $f_o \in \Phi$, and find an open set $U \subset X$ such that supp $f_o \subset U$ and U^- is compact. By the Urysohn lemma there is $\varphi \in \mathfrak{F}$ such that $0 \leq \varphi \leq 1$, $\varphi(x) = 1$ for $x \in$ supp f_o, and $\varphi(x) = 0$ for $x \in X - U$. According to (3.14), given $\varepsilon > 0$, we can find an $f_\varepsilon \in \Phi$ for which $f_\varepsilon < \varepsilon$. Since Φ is directed downwards, there is a $g \in \Phi$ such that $g \leq f_o \wedge f_\varepsilon$. Clearly, $g \leq \varepsilon\varphi$, and so by (4.3), $Ig \leq \varepsilon I\varphi$. It follows from the arbitrariness of ε that $I\Phi \searrow 0$. ///

We note that nonnegative linear functionals on $C_o(X)$, where X is a locally compact Hausdorff space, are sometimes called Radon measures in X (see, e.g., Ref. 5, Chapter III or Ref. 13, (9.1), p. 114).

For the rest of this chapter we shall assume that I^* and $I^\#$ are, respectively, Daniell and Bourbaki integrals on $\mathfrak{F}$. Our task will be to extend I^* to $\bar{\mathfrak{F}}^*$ and $I^\#$ to $\bar{\mathfrak{F}}^\#$.

(4.8) If $\{f_n\} \subset \mathfrak{F}$ is monotonic, $f \in \mathfrak{F}$, and $f_n \to f$, then $I^*f_n \to I^*f$.

(4.9) If $\Phi \subset \mathfrak{F}$, $f \in \mathfrak{F}$, and $\Phi \to f$, then $I^\#\Phi \to I^\#f$.

Since the proofs of (4.8) and (4.9) are analogous, we shall prove only (4.9), which is the harder of the two, and leave the proof of (4.8) to the reader.

Proof of (4.9). Suppose, for example, $\Phi \nearrow f$, and let $\Psi = \{f - g : g \in \Phi\}$. Then $\Psi \subset \mathfrak{F}$ and $\Psi \searrow 0$. Thus by the definition of the Bourbaki integral,

$$0 = \inf\{I^\#f - I^\#g : g \in \Phi\}$$
$$= I^\#f - \sup\{I^\#g : g \in \Phi\}$$

It follows that $I^\#\Phi \nearrow I^\#f$. ///

If $f \in \mathfrak{F}^*$ then there is a monotone sequence $\{f_n\} \subset \mathfrak{F}$ such that $f_n \to f$. Since $\{I^*f_n\}$ is a monotone sequence of real numbers, it has a limit in $\mathbf{R}^-$. In view of (4.8), it seems natural to extend the Daniell integral I^* by letting $I^*f = \lim I^*f_n$. In order to turn this idea into a legitimate definition, we first have to show that $\lim If_n$ depends only on f and not on the choice of $\{f_n\}$.

The setting for the extension of the Bourbaki integral $I^\#$ is the same; only monotone sequences have to be replaced by directed systems.

(4.10) Lemma. Let $\{f_n\}$ and $\{g_n\}$ be monotone sequences from $\mathfrak{F}$, and let $\lim f_n = f$ and $\lim g_n = g$. If $f \leq g$, then

$$\lim I^*f_n \leq \lim I^*g_n$$

Proof. Let $\lim I^*f_n = a$ and $\lim I^*g_n = b$. We have to distinguish four cases.

(a) Let $f_n \nearrow f$ and $g_n \searrow g$. Then $f_n \leq g_n$, and hence $I^*f_n \leq I^*g_n$, $n = 1, 2, \ldots$. Consequently, $a \leq b$.

(b) Let $f_n \nearrow f$ and $g_n \nearrow g$. Because $\lim_{n\to+\infty} (f_k - g_n) = f_k - g \leq 0$, we have $(f_k - g_n)^+ \searrow 0$ as $n \to +\infty$. Hence

$$0 = \lim_{n\to+\infty} I^*(f_k - g_n)^+ \geq \lim_{n\to+\infty} I^*(f_k - g_n)$$

$$= \lim_{n\to+\infty} (I^*f_k - I^*g_n) = I^*f_k - b$$

and so $I^*f_k \leq b$, $k = 1, 2, \ldots$. Thus again $a \leq b$.

(c) Let $f_n \searrow f$ and $g_n \searrow g$. Using $-f$ and $-g$ instead of g and f, respectively, this case is reduced to case (b).

(d) Let $f_n \searrow f$ and $g_n \nearrow g$. Then $f < +\infty$, $g > -\infty$, $a < +\infty$, $b > -\infty$, and so $f - g$ and $a - b$ have meaning. Because

$$\lim(f_n - g_n) = f - g \leq 0$$

we have $(f_n - g_n)^+ \searrow 0$. Hence,

$$0 = \lim I^*(f_n - g_n)^+ \geq \lim I^*(f_n - g_n)$$
$$= \lim(I^*f_n - I^*g_n) = a - b$$

and so $a \leq b$. ///

(4.11) Corollary. Let $\{f_n\}$ and $\{g_n\}$ be monotone sequences from $\mathfrak{F}$. Then

$$\lim I^*f_n = \lim I^*g_n$$

whenever $\lim f_n = \lim g_n$.

(4.12) Lemma. Let Φ and Ψ be directed systems from $\mathfrak{F}$, and let $\lim \Phi = f$ and $\lim \Psi = g$. If $f \leq g$, then

$$\lim I^{\#}\Phi \leq \lim I^{\#}\Psi$$

Proof. This proof resembles closely the proof of Lemma (4.10). Let $\lim I^{\#}\Phi = a$ and $\lim I^{\#}\Psi = b$. We shall distinguish four cases.

(a) Let $\Phi \nearrow f$ and $\Psi \searrow g$. Then $\varphi \leq \psi$ for each $\varphi \in \Phi$ and each $\psi \in \Psi$. Thus also $I^{\#}\varphi \leq I^{\#}\psi$ for all $\varphi \in \Phi$ and $\psi \in \Psi$. Therefore $a \leq b$.

(b) Let $\Phi \nearrow f$ and $\Psi \nearrow g$. Choose $\varphi \in \Phi$. Because

$$\inf\{\varphi - \psi : \psi \in \Psi\} = \varphi - g \leq 0$$

we have $\{(\varphi - \psi)^+ : \psi \in \Psi\} \searrow 0$. Thus,

$$0 = \inf\{I^{\#}(\varphi - \psi)^+ : \psi \in \Psi\} \geq \inf\{I^{\#}\varphi - I^{\#}\psi : \psi \in \Psi\}$$
$$= I^{\#}\varphi - \sup\{I^{\#}\psi : \psi \in \Psi\} = I^{\#}\varphi - b$$

and so $I^{\#}\varphi \leq b$. It follows from the arbitrariness of φ that $a \leq b$.

(c) Let $\Phi \searrow f$ and $\Psi \searrow g$. Using $-f$ and $-g$ instead of g and f, respectively, this case is reduced to case (b).

(d) Let $\Phi \searrow f$ and $\Psi \nearrow g$. Then f - g and a - b have meaning. Because

$$\inf\{\varphi - \psi : \varphi \in \Phi,\ \psi \in \Psi\} = f - g \leq 0$$

we have $\{(\varphi - \psi)^+ : \varphi \in \Phi,\ \psi \in \Psi\} \searrow 0$. Thus

$$\begin{aligned} 0 &= \inf\{I^{\#}(\varphi - \psi)^+ : \varphi \in \Phi,\ \psi \in \Psi\} \\ &\geq \inf\{I^{\#}\varphi - I^{\#}\psi : \varphi \in \Phi,\ \psi \in \Psi\} = a - b \end{aligned}$$

and so again $a \leq b$. ///

(4.13) Corollary. Let Φ and Ψ be directed systems from $\mathfrak{F}$. Then

$$\lim I^{\#}\Phi = \lim I^{\#}\Psi$$

whenever $\lim \Phi = \lim \Psi$.

(4.14) Definition. Given $f \in \mathfrak{F}^*$ and a monotone sequence $\{f_n\} \subset \mathfrak{F}$ such that $f_n \to f$, we let

$$I^*f = \lim I^*f_n$$

(4.15) Definition. Given $f \in \mathfrak{F}^{\#}$ and a directed system $\Phi \subset \mathfrak{F}$ such that $\Phi \to f$, we let

$$I^{\#}f = \lim I^{\#}\Phi$$

By Corollaries (4.11) and (4.13), the previous definitions give well-defined maps

$$I^* : \bar{\mathfrak{F}}^* \to R^- \qquad \text{and} \qquad I^{\#} : \bar{\mathfrak{F}}^{\#} \to R^-$$

It follows from (4.8) and (4.9) that these maps are extensions of the Daniell and Bourbaki integrals

$$I^* : \mathfrak{F} \to \mathbf{R} \qquad \text{and} \qquad I^{\#} : \mathfrak{F} \to \mathbf{R}$$

Therefore, no confusion can arise if these extended integrals are denoted by the same symbols as the original ones.

The following are easy consequences of Definitions (4.14) and (4.15) and Lemmas (4.10) and (4.12).

(4.16) If $f \in \mathfrak{F}_+^*$, then

$$I^*f = \sup\{I^*g : g \in \mathfrak{F},\ g \le f\}$$

and if $f \in \mathfrak{F}_-^*$, then

$$I^*f = \inf\{I^*g : g \in \mathfrak{F},\ g \ge f\}$$

(4.17) If $f \in \mathfrak{F}_+^\#$, then

$$I^\#f = \sup\{I^\#g : g \in \mathfrak{F},\ g \le f\}$$

and if $f \in \mathfrak{F}_-^\#$, then

$$I^\#f = \inf\{I^\#g : g \in \mathfrak{F},\ g \ge f\}$$

(4.18) $$(f,g \in \bar{\mathfrak{F}}^*,\ f \le g) \Rightarrow I^*f \le I^*g$$

(4.19) $$(f,g \in \bar{\mathfrak{F}}^\#,\ f \le g) \Rightarrow I^\#f \le I^\#g$$

(4.20) If $f \in \bar{\mathfrak{F}}^*$ and $a \in \mathbf{R}$, then $af \in \bar{\mathfrak{F}}^*$ and $I^*(af) = aI^*f$.

(4.21) If $f \in \bar{\mathfrak{F}}^\#$ and $a \in \mathbf{R}$, then $af \in \bar{\mathfrak{F}}^\#$ and $I^\#(af) = aI^\#f$.

(4.22) If $f,g \in \mathfrak{F}_+^*$ or $f,g \in \mathfrak{F}_-^*$, then

$$I^*(f+g) = I^*f + I^*g$$

(4.23) If $f,g \in \mathfrak{F}_+^\#$ or $f,g \in \mathfrak{F}_-^\#$, then

$$I^\#(f+g) = I^\#f + I^\#g$$

The proofs of (4.16)-(4.23) are very simple. As a sample we shall prove (4.23). The proofs of the remaining statements are left to the reader.

Proof of (4.23). Let $f,g \in \mathfrak{F}_+^\#$, and let

$$\Gamma = \{\varphi + \psi : \varphi,\psi \in \mathfrak{F},\ \varphi \le f,\ \psi \le g\}$$

It follows from (2.16) that $\Gamma \nearrow f + g$. Hence,

$$
\begin{aligned}
I^{\#}(f+g) &= \sup\{I^{\#}\gamma : \gamma \in \Gamma\} \\
&= \sup\{I^{\#}\varphi : \varphi \in \mathfrak{F},\ \varphi \le f\} + \sup\{I^{\#}\psi : \psi \in \mathfrak{F},\ \psi \le g\} \\
&= I^{\#}f + I^{\#}g \quad ///
\end{aligned}
$$

In view of (2.10) and (2.18) one would naturally expect the following two propositions.

(4.24) Proposition. Let $\{f_n\} \subset \mathfrak{F}_+^*$. If $f_n \nearrow f$ then $I^*f_n \nearrow I^*f$.

Proof. By (2.10), $f \in \mathfrak{F}_+^*$, and so I^*f is defined. Moreover, there is a sequence $\{g_n\} \subset \mathfrak{F}$ such that $g_n \le f_n$, $n = 1, 2, \ldots$, and $g_n \nearrow f$. Thus by (4.18),

$$I^*g_n \le I^*f_n \le I^*f \qquad (n = 1, 2, \ldots)$$

Passing to the limit we obtain

$$I^*f = \lim I^*g_n \le \lim I^*f_n \le I^*f \quad ///$$

(4.25) Proposition. Let $\Phi \subset \mathfrak{F}_+^{\#}$. If $\Phi \nearrow f$, then $I^{\#}\Phi \nearrow I^{\#}f$.

Proof. By (2.18), $f \in \mathfrak{F}_+^{\#}$, and so $I^{\#}f$ is defined. Moreover, if $\Psi = \bigcup_{g\in\Phi}\{h \in \mathfrak{F} : h \le g\}$, then $\Psi \nearrow f$. Hence, using (4.17), we obtain

$$
\begin{aligned}
I^{\#}f &= \sup\{I^{\#}h : h \in \Psi\} \\
&= \sup\{\sup\{I^{\#}h : h \in \mathfrak{F},\ h \le g\} : g \in \Phi\} \\
&= \sup\{I^{\#}g : g \in \Phi\} \quad ///
\end{aligned}
$$

(4.26) Remark. In view of (4.16) it is obvious that I^* can also be extended from $\mathfrak{F}$ to $\bar{\mathfrak{F}}^*$ by letting

$$I^*f = \sup\{I^*g : g \in \mathfrak{F},\ g \le f\}$$

for $f \in \mathfrak{F}_+^*$ and

$$I^*f = \inf\{I^*g : g \in \mathfrak{F},\ g \ge f\}$$

for $f \in \mathfrak{F}_-^*$. Using this approach we do not need Lemma (4.10), but we have to work harder to prove (4.18) and (4.24).

A similar remark applies to the extension of $I^{\#}$.

Exercises

In exercises (4-1) through (4-3), X denotes an arbitrary set.

(4-1)[+] Let $\mathfrak{F}$ be any fundamental system on X, and let $x_o \in X$. For $f \in \mathfrak{F}$, let $If = f(x_o)$. This functional I is called the Dirac integral at x_o. Show that I is a Bourbaki integral on $\mathfrak{F}$ and that $If = f(x_o)$ for each $f \in \bar{\mathfrak{F}}^{\#}$.

(4-2) Assume that X is an infinite set, and let $\mathfrak{F}$ consist of all real-valued functions f on X for which there is a $c_f \in \mathbf{R}$ such that the set $\{x \in X : f(x) \neq c_f\}$ is finite [see (2-4)]. For $f \in \mathfrak{F}$, let $If = c_f$. Show that

(i) I is a nonnegative linear functional on $\mathfrak{F}$.

(ii) I is not a Bourbaki integral.

(iii) I is a Daniell integral iff X is uncountable.

(4-3) Let $\aleph$ be an infinite cardinal, and let $\|X\| > \aleph$. Suppose that $\mathfrak{F}$ consists of all real-valued functions on X for which there is a $c_f \in \mathbf{R}$ such that

$$\|\{x \in X : f(x) \neq c_f\}\| \leq \aleph$$

[see (2-5)]. For $f \in \mathfrak{F}$, let $If = c_f$.

(i) Show that I is a Daniell but not a Bourbaki integral on $\mathfrak{F}$.

(ii) Observe that if $X = W$ or $X = W^-$ [see Chapter 1, Section B] and $\aleph = \aleph_o$, then the system $\mathcal{C}(X)$ is an example of the fundamental system $\mathfrak{F}$ from this exercise. Hint: See (1-15)(ii).

(4-4)[+] Let X be a discrete space, and let $\mathfrak{F} = \mathcal{C}_o(X)$ [see (3-1) and (2-6)]. For $f \in \mathfrak{F}$, set

$$If = \Sigma_{x \in X} f(x)$$

Show that the sum on the right side has meaning and that I is a Bourbaki integral on $\mathfrak{F}$.

(4-5)[+] Let $X = [0,1]$, and let $\mathfrak{F}_1$ consist of all continuous functions f on X for which $f(0) = 0$ and $f'_+(0)$ exists [see (2-8)(a)]. For $f \in \mathfrak{F}_1$, let $If = f'_+(0)$, and show that

(i) I is a nonnegative linear functional on $\mathfrak{F}_1$.

(ii) I is not a Daniell integral on $\mathfrak{F}_1$.

(iii) I is a Bourbaki integral on the system $\mathfrak{F}_2$ consisting of all linear functions from $\mathfrak{F}_1$ [see (2-8)(b)].

In exercises (4-6) and (4-7) we shall use the Riemann-Stieltjes integral $\int_a^b f(x)\, dg(x)$ (see Ref. 1, Chapter 9). The reader unfamiliar with the Riemann-Stieltjes integral should work these exercises for the Riemann integral $\int_a^b f(x)\, dx$.

(4-6)$^+$ Let $a, b \in \mathbf{R}$, $a < b$, $X = [a,b]$, $\mathfrak{F} = C(X)$, and let g be a nondecreasing real-valued function on X. For $f \in \mathfrak{F}$, let

$$If = \int_a^b f(x)\, dg(x)$$

Show that I is a Bourbaki integral on $\mathfrak{F}$.

(4-7)$^+$ Let $X = \mathbf{R}$, $\mathfrak{F} = C_o(X)$, and let g be a nondecreasing real-valued function on X. For $f \in \mathfrak{F}$, let

$$If = \lim_{a \to +\infty} \int_{-a}^{a} f(x)\, dg(x)$$

Show that limit on the right side exists and that I is a Bourbaki integral on $\mathfrak{F}$.

(4-8)* Let $X = \mathbf{R}$, and let $\mathfrak{F}$ consist of all finite continuous functions on X vanishing exponentially at infinity [see (2-9)(vii)]. For $f \in \mathfrak{F}$, let

$$If = \lim_{a \to +\infty} \int_{-a}^{a} f(x)\, dx$$

Show that the limit on the right side exists and that I is a Bourbaki integral on $\mathfrak{F}$.

(4-9)* Let $X = \mathbf{R}$ and let $\mathfrak{F}$ consist of all real-valued continuous functions on X of polynomial growth [see (2-9)(v)]. For $f \in \mathfrak{F}$, let

$$If = \lim_{a \to +\infty} \int_{-a}^{a} e^{-x^2} f(x)\, dx$$

Show that the limit on the right side exists and that I is a Bourbaki integral on $\mathfrak{F}$.

(4-10)⁺ Let $X = \mathbf{R}^m$, $m \geq 1$ an integer, and let $\mathfrak{F} = C_o(X)$. For $f \in \mathfrak{F}$, let

$$If = \int_K f(x)\, dx$$

where K is a compact interval containing supp f and $\int_K f(x)\, dx$ is the multiple Riemann integral of f on K (see Ref. 1, Chapter 10). Show that the right side is independent of the choice of K and that I is a Bourbaki integral on $\mathfrak{F}$.

(4-11) Let $\mathfrak{F}$ be a fundamental system on a set X, and let $\{I_n\}$ be a sequence of nonnegative linear functionals on $\mathfrak{F}$. Suppose that the series $\Sigma_{n=1}^{\infty} I_n f$ converges for all $f \in \mathfrak{F}$, and set

$$If = \sum_{n=1}^{\infty} I_n f$$

for each $f \in \mathfrak{F}$. Show that

(i) I is a nonnegative linear functional on $\mathfrak{F}$.
(ii) I is a Daniell integral on $\mathfrak{F}$ iff each I_n is.
(iii) I is a Bourbaki integral on $\mathfrak{F}$ iff each I_n is.

(4-12) Let $\mathfrak{F}$ be a fundamental system on a set X such that all functions from $\mathfrak{F}$ are <u>bounded</u>. Let $\{x_n\} \subset X$, and let $\{a_n\}$ be a sequence of positive real numbers. Assume that $\Sigma_{n=1}^{\infty} a_n$ converges, and set

$$If = \sum_{n=1}^{\infty} a_n f(x_n)$$

for each $f \in \mathfrak{F}$. Show that I is a Bourbaki integral on $\mathfrak{F}$. Hint: Use (4-1) and (4-11).

(4-13) Let X be a locally compact Hausdorff space, and let $\mathfrak{F} = C_0(X)$. We say that a net $\{f_\alpha\} \subset \mathfrak{F}$ (see Ref. 16, Chapter 2) converges to $f \in \mathfrak{F}$ whenever there is a compact set $C \subset X$ such that eventually $\operatorname{supp} f_\alpha \subset C$ and for every $\varepsilon > 0$ eventually $|f_\alpha - f| < \varepsilon$. Show that this convergence defines in $\mathfrak{F}$ a Hausdorff topology, and prove that in this topology every nonnegative linear functional on $\mathfrak{F}$ is continuous.

(4-14)$^+$ Let X be a locally compact Hausdorff space, and let I be a linear functional on $\mathfrak{F} = C_0(X)$. Denote by supp I the set of all points $x \in X$ with the following property: For every neighborhood U of x there is a function $f \in \mathfrak{F}$ such that $\operatorname{supp} f \subset U$ and $If \neq 0$. The set supp I is called the support of the functional I. Show that

(i) The set supp I is closed. Hint: Observe that the complement of supp I is open.

(ii) If $f \in \mathfrak{F}$ and $\operatorname{supp} f \cap \operatorname{supp} I = \emptyset$, then $If = 0$. Hint: Use a partition of unity (see Ref. 16, Chapter 5, problem W, p. 171) to show that $f = \Sigma_{i=1}^n f_n$ with $If_i = 0$, $i = 1, \ldots, n$.

(iii) Let I be a nonnegative linear functional on $\mathfrak{F}$, and let $f \in \mathfrak{F}$ be a nonnegative function. If $f(x) > 0$ for some $x \in \operatorname{supp} I$, then $If > 0$.

5. UPPER AND LOWER INTEGRALS; SUMMABLE FUNCTIONS

Much of the following exposition holds equally well for both Daniell and Bourbaki integrals. Therefore, whenever we do not have to distinguish between I^* and $I^\#$, we shall merely write I and call this symbol an integral. Similarly, $\mathfrak{F}_+$, $\mathfrak{F}_-$, and $\bar{\mathfrak{F}}$ will denote both $\mathfrak{F}_+^*$ or $\mathfrak{F}_+^\#$, $\mathfrak{F}_-^*$ or $\mathfrak{F}_-^\#$, and $\bar{\mathfrak{F}}^*$ or $\bar{\mathfrak{F}}^\#$, respectively. We shall adhere to this convention throughout the book.

A heuristic motivation for what follows comes from analyzing the definition of the upper Riemann integral.

Let $P = \{a = x_0 < \cdots < x_n = b\}$ be a partition of a bounded interval $[a,b)$, and let f be a bounded function on $[a,b)$. Let

$$S(P) = \sum_{i=1}^{n} M_i(x_i - x_{i-1})$$

where $M_i = \sup\{f(x) : x \in [x_{i-1}, x_i)\}$, $i = 1, \ldots, n$. The <u>upper Riemann integral</u> of f on $[a,b)$ is classically defined by the formula

$$\bar{\int_a^b} f(x)\,dx = \inf S(P)$$

where the infimum is taken over all partitions P of $[a,b)$. We shall interpret this definition in terms of step functions [see (2-11)]. A function g on $[a,b)$ is called a <u>step function</u> if there is a partition $P = \{a = x_0 < \cdots < x_n = b\}$ of $[a,b)$ and real numbers $M_1, \ldots, M_n$ such that $g(x) = M_i$ for all $x \in [x_{i-1}, x_i)$. Clearly, g is Riemann integrable and

$$\int_a^b g(x)\,dx = \sum_{i=1}^{n} M_i(x_i - x_{i-1})$$

Thus we see that

$$\overline{\int_a^b} f(x)\,dx = \inf \int_a^b g(x)\,dx$$

where the infimum is taken over all step functions $g \geq f$.

We shall proceed analogously and define an upper integral by the formula

$$\overline{I}f = \inf Ig$$

where the infimum is taken over all $g \in \mathfrak{F}_+$ for which $g \geq f$. We shall see that this, seemingly small, change in definition will yield an integral with much better properties than those of the Riemann integral.

Throughout this chapter we shall assume that X is an arbitrary set, $\mathfrak{F}$ is a fundamental system on X, and that I is an integral on $\mathfrak{F}$ which has been extended to $\overline{\mathfrak{F}}$.

(5.1) <u>Definition</u>. Let f be a function on X. The extended real numbers

$$\overline{I}f = \inf\{Ig : g \in \mathfrak{F}_+ ,\ g \geq f\}$$

and

$$\underline{I}f = \sup\{Ig : g \in \mathfrak{F}_- ,\ g \leq f\}$$

are called the *upper* and *lower integrals* of f, respectively.

From the definition we obtain the following:

(5.2) $$\underline{I}f \leq \overline{I}f \quad \text{and} \quad \underline{I}f = -\overline{I}(-f)$$

(5.3) $\overline{I}(af) = a\overline{I}f$ and $\underline{I}(af) = a\underline{I}f$ for all $a \in \mathbb{R}$, $a \geq 0$.

(5.4) $\overline{I}f \leq \overline{I}g$ and $\underline{I}f \leq \underline{I}g$ whenever $f \leq g$.

The next claim requires a short proof.

(5.5) $$f \in \overline{\mathfrak{F}} \Rightarrow \overline{I}f = \underline{I}f = If$$

Proof. With no loss of generality we may assume that $f \in \mathfrak{F}_+$. Then clearly, $\bar{I}f = If$. Since $\mathfrak{F} \subset \mathfrak{F}_+ \cap \mathfrak{F}_-$, it follows from (4.16) and (4.17) that

$$\underline{I}f \leq \bar{I}f = If = \sup\{Ig : g \in \mathfrak{F},\ g \leq f\} \leq \underline{I}f \quad ///$$

(5.6) Proposition. Let f, g, and h be functions on X, and let

$$h(x) = f(x) + g(x)$$

for each $x \in X$ for which $f(x) + g(x)$ has meaning. Then

$$\bar{I}h \leq \bar{I}f + \bar{I}g$$

whenever $\bar{I}f + \bar{I}g$ has meaning.

Proof. Suppose that $\bar{I}f + \bar{I}g$ has meaning and that

$$\bar{I}f + \bar{I}g < \bar{I}h$$

Then there are $\varphi, \psi \in \mathfrak{F}_+$ such that $\varphi \geq f$, $\psi \geq g$, and

$$I(\varphi + \psi) = I\varphi + I\psi < \bar{I}h$$

If $f(x) + g(x)$ has meaning, then

$$\varphi(x) + \psi(x) \geq f(x) + g(x) = h(x)$$

If $f(x) + g(x)$ has no meaning, then either $f(x) = +\infty$ or $g(x) = +\infty$. It follows that

$$\varphi(x) + \psi(x) = +\infty \geq h(x)$$

Thus $\varphi + \psi \geq h$, and because $\varphi + \psi \in \mathfrak{F}_+$ [see (2.7) and (2.15)], we have

$$\bar{I}h \leq I(\varphi + \psi)$$

a contradiction. ///

(5.7) Remark. There is a certain peculiarity in Proposition (5.6). Namely, if $f(x) + g(x)$ has meaning for no $x \in X$, then h is quite arbitrary. Yet, if $\bar{I}f + \bar{I}g$ has meaning, we still have

$$\bar{I}h \leq \bar{I}f + \bar{I}g$$

A full explanation of this, seemingly paradoxical, situation will be given later [see (11.6)]. Roughly speaking, it will turn out that if $\bar{I}f + \bar{I}g$ has meaning, then $f(x) + g(x)$ has meaning for "almost all" $x \in X$. The expression "almost all" will be defined precisely in (11.1). The reader can, however, get some feeling for what is going on from the next proposition.

(5.8) Proposition. Let f and g be functions on X, and let $\bar{I}f < +\infty$. If $g(x) = f(x)$ for each $x \in X$ for which $f(x) < +\infty$, then $\bar{I}g = \bar{I}f$.

Proof. For $x \in X$, let $h(x) = 0$ if $f(x) < +\infty$, and $h(x) = +\infty$ if $f(x) = +\infty$. Because $\bar{I}f < +\infty$, there is a $\varphi \in \mathfrak{F}_+$ such that $\varphi \geq f$ and $I\varphi$ is finite. Clearly, $h \geq 0$ and

$$h(x) = \varphi(x) - \varphi(x)$$

for each $x \in X$ for which $\varphi(x) - \varphi(x)$ has meaning. Thus by (5.6), (5.2), and (5.5),

$$0 \leq \bar{I}h \leq I\varphi - I\varphi = 0$$

We also have $g \leq f$ and

$$f(x) = g(x) + h(x)$$

for each $x \in X$ for which $g(x) + h(x)$ has meaning. Thus by (5.4) and (5.6)

$$\bar{I}g \leq \bar{I}f \leq \bar{I}g + \bar{I}h = \bar{I}g \quad ///$$

The following theorem is of paramount importance to the whole theory of integration. In the literature it is usually referred to as the "monotone convergence theorem."

(5.9) Theorem. Let f and f_n, $n = 1, 2, \ldots$, be functions on X. If $f_n \nearrow f$ and $\bar{I}f_1 > -\infty$, then $\bar{I}f_n \nearrow \bar{I}f$.

Proof. Clearly, $\lim \bar{I}f_n \leq \bar{I}f$, and so the theorem is true if $\lim \bar{I}f_n = +\infty$. Hence assume that $\lim \bar{I}f_n < +\infty$. Since $\bar{I}f_1 > -\infty$, it follows that $\bar{I}f_n \neq \pm\infty$, $n = 1, 2, \ldots$. Choose $\epsilon > 0$ and $g_n \in \mathfrak{F}_+$ such that $g_n \geq f_n$ and

$$Ig_n < \bar{I}f_n + \epsilon 2^{-n} \qquad (*)$$

Letting $h_n = \vee_{i=1}^n g_i$, we have $h_n \in \mathfrak{F}_+$ [see (2.7) and (2.15)] and $h_n \nearrow h$ for some function $h \in \mathfrak{F}_+$ [see (2.10) and (2.18)]. Since Since $h_n \geq g_n \geq f_n$, $n = 1, 2, \ldots$, we obtain $h \geq f$ and consequently $\bar{I}f \leq Ih$. Suppose we have proved that

$$Ih_n < \bar{I}f_n + \Sigma_{i=1}^n \epsilon 2^{-i} \qquad (**)$$

$n = 1, 2, \ldots$. Then by (4.24) and (4.25),

$$\bar{I}f \leq Ih = \lim Ih_n \leq \lim \bar{I}f_n + \epsilon$$

and the theorem would follow from the arbitrariness of ϵ. Thus it remains to prove (**), which we shall do by induction. For $n = 1$, inequality (**) reduces to (*) as $h_1 = g_1$. Assume that (**) is correct for some $n \geq 1$. Since $f_n \leq h_n \wedge g_{n+1}$ and

$$h_{n+1} + h_n \wedge g_{n+1} = h_n \vee g_{n+1} + h_n \wedge g_{n+1} = h_n + g_{n+1}$$

we have $\bar{I}f_n \leq I(h_n \wedge g_{n+1})$ and

$$Ih_{n+1} + I(h_n \wedge g_{n+1}) = Ih_n + Ig_{n+1}$$

From this, the induction hypothesis, and (*) it follows that

$$\begin{aligned} Ih_{n+1} &\leq Ih_n + Ig_{n+1} - \bar{I}f_n \\ &\leq \bar{I}f_n + \Sigma_{i=1}^n \epsilon 2^{-i} + \bar{I}f_{n+1} + \epsilon 2^{-n-1} - \bar{I}f_n \\ &= If_{n+1} + \Sigma_{i=1}^{n+1} \epsilon 2^{-i} \quad /// \end{aligned}$$

We note that Theorem (5.9) is generally false if the assumption $\bar{I}f_1 > -\infty$ is omitted [see (5-10)(iv)].

(5.10) Proposition (Fatou). Let f_n, $n = 1, 2, \ldots$, be functions on X. If $\bar{I}(\inf f_n) > -\infty$, then

$$\bar{I}(\inf f_n) \leq \bar{I}(\liminf f_n) \leq \liminf \bar{I}f_n$$

Proof. Let $g_n = \inf\{f_k : k \geq n\}$, $n = 1, 2, \ldots$. Then $g_n \leq f_n$, $\bar{I}g_1 > -\infty$, and $g_n \nearrow \liminf f_n$. Thus by (5.4) and (5.9),

$$\bar{I}(\inf f_n) \leq \bar{I}(\liminf f_n) = \lim \bar{I}g_n \leq \liminf \bar{I}f_n \quad ///$$

In the literature Proposition (5.10) is frequently referred to as "Fatou's lemma."

Obviously, there are dual statements to (5.6)-(5.10). We shall leave their formulation and proofs to the reader.

The next definition is analogous to the definition of the Riemann integral.

(5.11) Definition. A function f on X is called summable whenever

$$\underline{I}f = \bar{I}f \neq \pm\infty$$

The family of all summable functions on X is denoted by $\mathfrak{L}$.

(5.12) Remark. A comment on the above mentioned analogy between the definitions of summable functions and Riemann integrable functions seems to be in place. The condition $\underline{I}f = \bar{I}f$ comes, indeed, from this analogy. However, the condition $\bar{I}f \neq \pm\infty$, which is automatically satisfied in the Riemann theory, is an additional requirement in our case. Although, we shall often use the finiteness requirement in proving some basic theorems [e.g., (5.15) and (5.17)], at this point, it is not clear why we cannot build the theory without it. An answer to this question will be given in Chapter 10 [see (10-11)].

According to (5.5), it is consistent to write

$$If = \bar{I}f$$

for $f \in \mathfrak{L}$. Thus we have already extended the original integral I on $\mathfrak{F}$ to a map

$$I : \mathfrak{L} \cup \bar{\mathfrak{F}} \to R^-$$

We shall see that the extension to $\mathcal{L}$ is most significant. It follows at once from (5.5) that

(5.13) $$(f \in \overline{\mathfrak{F}} \text{ and } If \neq \pm\infty) \Rightarrow f \in \mathcal{L}$$

From (5.3) we obtain the following.

(5.14) Theorem. If $f \in \mathcal{L}$ and $a \in \mathbf{R}$, then $af \in \mathcal{L}$ and

$$I(af) = aIf$$

(5.15) Theorem. Let $f, g \in \mathcal{L}$, and let h be a function on X. If

$$h(x) = f(x) + g(x)$$

for each $x \in X$ for which $f(x) + g(x)$ has meaning, then $h \in \mathcal{L}$ and

$$Ih = If + Ig$$

Proof. From (5.6), its dual, and (5.2) we obtain

$$-\infty < If + Ig \leq \underline{I}h \leq \overline{I}h \leq If + Ig < +\infty \quad ///$$

Theorems (5.14) and (5.15) imply that $\mathcal{L}$ is "almost" a vector space and that I is a linear functional on $\mathcal{L}$. The only missing property is the unique definition of the sum $f + g$ for $f, g \in \mathcal{L}$. Thus, e.g., the system $\mathcal{L}_o$ of all finite functions from $\mathcal{L}$ forms a genuine vector space over $\mathbf{R}$. The next proposition indicates that, in a sense, $\mathcal{L}_o$ is not much smaller than $\mathcal{L}$.

(5.16) Proposition. Let $f \in \mathcal{L}$, and let g be a function on X such that $g(x) = f(x)$ for each $x \in X$ for which $f(x)$ is finite. Then $g \in \mathcal{L}$ and $Ig = If$.

This proposition is an easy corollary of (5.8) and its dual.

(5.17) Theorem (Lebesgue). Let $\{f_n\} \subset \mathcal{L}$, $g, h \in \mathcal{L}$, and let $g \leq f_n \leq h$, $n = 1, 2, \ldots$. If $f = \lim f_n$, then $f \in \mathcal{L}$ and

$$If = \lim If_n$$

Proof. Since

$$g \leq \inf f_n \leq \sup f_n \leq h$$

we have $\bar{I}(\inf f_n) > -\infty$ and $\underline{I}(\sup f_n) < +\infty$. An application of (5.10) and its dual gives

$$-\infty < \bar{I}(\inf f_n) \leq \bar{I}(\liminf f_n)$$
$$= \bar{I}f \leq \liminf If_n \leq \limsup If_n$$
$$\leq \underline{I}(\limsup f_n) = \underline{I}f \leq \underline{I}(\sup f_n) < +\infty$$

The theorem follows from (5.2). ///

In the literature Theorem (5.17) is often referred to as the "Lebesgue dominated convergence theorem" or simply as the "dominated convergence theorem."

(5.18) Lemma. Let $f \in \mathfrak{L}$ and $\epsilon > 0$. Then there is a $g \in \mathfrak{F}$ such that $\bar{I}(|f - g|) < \epsilon$.

Proof. Choose an $h \in \mathfrak{F}_+$ so that $h \geq f$ and $Ih < If + \epsilon/2$. By the definition of I on $\mathfrak{F}_+$, there is a $g \in \mathfrak{F}$ such that $g \leq h$ and $Ig > Ih - \epsilon/2$. For $x \in X$, let

$$k(x) = h(x) - f(x)$$

if $h(x) - f(x)$ has meaning and $k(x) = +\infty$ otherwise. Thus $k \in \mathfrak{L}$, $k \geq 0$, and

$$Ik = Ih - If < \epsilon/2$$

Moreover,

$$|f - g| \leq k + (h - g)$$

Indeed, this is a triangle inequality if $h(x) - f(x)$ has meaning; otherwise, $k(x) = +\infty$ and the inequality holds trivially. Thus,

$$\bar{I}(|f - g|) \leq Ik + Ih - Ig < \epsilon \quad ///$$

(5.19) Theorem. If $f \in \mathfrak{L}$, then also $|f| \in \mathfrak{L}$.

Proof. Let $f \in \mathfrak{L}$, and let $\epsilon > 0$. By (5.18), there is a $g \in \mathfrak{F}$ such that $\bar{I}(|f - g|) < \epsilon$. Since $|g| \in \mathfrak{F}$ and

$$|g| - |f - g| \leq |f| \leq |g| + |f - g|$$

(5.6) and its dual yield

$$\begin{aligned} I(|g|) - \varepsilon < I(|g|) - \bar{I}(|f - g|) \\ &= I(|g|) + \underline{I}(-|f - g|) \leq \underline{I}(|f|) \leq \bar{I}(|f|) \\ &\leq I(|g|) + \bar{I}(|f - g|) < I(|g|) + \varepsilon \end{aligned}$$

Thus

$$0 \leq \bar{I}(|f|) - \underline{I}(|f|) < 2\varepsilon$$

and the theorem follows from the arbitrariness of ε. ///

By analogy with infinite series, the previous theorem is usually abbreviated as "the integral I is absolutely convergent."

Using Theorem (5.19), we can reformulate Lemma (5.18) as follows.

(5.20) Corollary. Let $f \in \mathfrak{L}$ and $\varepsilon > 0$. Then there is a $g \in \mathfrak{F}$ such that $I(|f - g|) < \varepsilon$.

Because

$$f(x) \vee g(x) = \frac{1}{2}[f(x) + g(x) + |f(x) - g(x)|]$$

$$f(x) \wedge g(x) = \frac{1}{2}[f(x) + g(x) - |f(x) - g(x)|]$$

for each $x \in X$ for which the right side has meaning, we also obtain:

(5.21) Corollary. If $f, g \in \mathfrak{L}$, then also $f \vee g \in \mathfrak{L}$ and $f \wedge g \in \mathfrak{L}$.

Since $f = f^+ - f^-$ for every function f on X, we have, in particular,

(5.22) Corollary. A function f on X belongs to $\mathfrak{L}$ iff both f^+ and f^- do.

(5.23) Remark. Theorems (5.14), (5.15), (5.19), and (5.17) imply the following noteworthy fact: The system $\mathfrak{L}_o$ of all real-valued functions from $\mathfrak{L}$ is a fundamental system on X, and I is a Daniell integral on $\mathfrak{L}_o$.

Exercises

If I is a Bourbaki integral on a fundamental system $\mathfrak{F}$, then by (4.6), I is also a Daniell integral on $\mathfrak{F}$. Thus I can be extended in two different ways: as a Daniell integral and as a Bourbaki integral. To distinguish between these two extensions we shall introduce the symbols I^*, $I^{\#}$, $\bar{I}^*$, $\bar{I}^{\#}$, $\underline{I}^*$, $\underline{I}^{\#}$, $\mathfrak{L}^*$, and $\mathfrak{L}^{\#}$, whose meanings are obvious.

In exercises (5-1) through (5-6) let $x_o \in X$, and let I be the Dirac integral at x_o [see (4-1)].

(5-1) Let X be an arbitrary set, and let $\mathfrak{F}$ be an arbitrary fundamental system on X. Show that

(i) $\underline{I}f \leq f(x_o) \leq \bar{I}f$ for every function f on X.

(ii) $f \in \mathfrak{L} \cup \bar{\mathfrak{F}} \Rightarrow If = f(x_o)$.

(5-2) Let X be an arbitrary set, and let $\mathfrak{F}$ consist of all bounded functions on X [see (2-1)]. Show that $\mathfrak{L}^* = \mathfrak{L}^{\#}$ consists of all functions f on X for which $f(x_o) \neq \pm\infty$.

(5-3) Let X be an arbitrary set, and let $\mathfrak{F}$ consist of all real-valued functions f on X such that $f(x) = 0$ for all $x \in X$, $x \neq x_o$. Show that

(i) $\mathfrak{F}$ is a fundamental system on X.

(ii) $\mathfrak{L}^* = \mathfrak{L}^{\#} = \mathfrak{F}$.

(iii) If a finite function f on X does not belong to $\mathfrak{F}$, then $\underline{I}f < \bar{I}f$ and either $\underline{I}f = -\infty$ or $\bar{I}f = +\infty$.

(5-4) Let X be a discrete space, and let $\mathfrak{F} = C_o(X)$ [see (3-1)]. Show that

(i) $\mathfrak{L}^*$ consists of all functions f on X such that $f(x_o) \neq \pm\infty$ and supp f is countable.

(ii) $\mathfrak{L}^{\#}$ consists of all functions f on X such that $f(x_o) \neq \pm\infty$.

(iii) If $f \in \mathfrak{L}^{\#} - \mathfrak{L}^*$, then $\underline{I}^*f < \bar{I}^*f$ and either $\underline{I}^*f = -\infty$ or $\bar{I}^*f = +\infty$.

(5-5)[+] Let $X = W^-$ [see Chapter 1, Section B], $x_o = \Omega$, and let $\mathfrak{F} = C(X)$. Show that

(i) $\mathcal{L}^*$ consists of all functions f on X which are eventually finite constants.

(ii) $\mathcal{L}^\#$ consists of all functions f on X such that $f(x_o) \neq \pm\infty$.

(iii) $f \in \mathcal{L}^\# - \mathcal{L}^* \Rightarrow \underline{I}^*f < \overline{I}^*f$.

(5-6) Let $X = [0,1]$, $x_o = 1$, and let $\mathfrak{F}$ consist of all linear functions f on X for which $f(0) = 0$ [see (2-8)(b)]. Show that $\mathcal{L}^* = \mathcal{L}^\# = \mathfrak{F}$ and that $\underline{I}f < \overline{I}f$ for every finite function f on X which does not belong to $\mathfrak{F}$.

(5-7) Let $\aleph$ be an infinite cardinal, and let $\|X\| > \aleph$. Suppose that $\mathfrak{F}$ consists of all real-valued functions f on X for which there is a $c_f \in \mathbf{R}$ such that

$$\|\{x \in X : f(x) \neq c_f\}\| \leq \aleph$$

[see (2-5)]. According to (4-3) we can define a Daniell integral I on $\mathfrak{F}$ by letting $If = c_f$ for all $f \in \mathfrak{F}$. Show that $\mathcal{L} = \mathfrak{F}$ and that $\underline{I}f < \overline{I}f$ for every finite function f on X which does not belong to $\mathfrak{F}$.

(5-8)+ Let A be an uncountable set of positive real numbers. Show that there is an $\varepsilon > 0$ such that the set $A_\varepsilon = \{t \in A : t > \varepsilon\}$ is also uncountable. Hint: Show that $A = \bigcup_{n=1}^{\infty} A_{1/n}$.

(5-9)+ Let X be an arbitrary set, and let $\mathfrak{J}$ be the family of all finite subsets of X. If f is a nonnegative function on X, let

$$\Sigma f = \sup\{\sum_{x \in F} f(x) : F \in \mathfrak{J}\}$$

If f is an arbitrary function on X, let

$$\Sigma f = \Sigma f^+ - \Sigma f^-$$

whenever the difference on the right side has meaning. Instead of Σf, we shall sometimes write $\Sigma_{x \in X} f(x)$ or $\Sigma\{f(x) : x \in X\}$.

Let f be a function on X for which Σf is defined. Prove

(i) $\Sigma f \neq \pm\infty$ iff $\Sigma|f| < +\infty$.

(ii) If $\Sigma f \neq \pm\infty$, then the set $\{x \in X : f(x) \neq 0\}$ is countable. Hint: Use (i) and (5-8).

(iii) If $\Sigma f \neq \pm\infty$ and $\{x_1, x_2, \ldots\} = \{x \in X : f(x) \neq 0\}$, then the series (possibly finite) $\Sigma_n f(x_n)$ is absolutely convergent and $\Sigma f = \Sigma_n f(x_n)$.

(iv) If $A \subset X$, then $\Sigma_{x \in A} f(x)$ is defined. Hint: Use (i).

(v) Let $\{A_\alpha : \alpha \in T\}$ be a disjoint family of sets such that $\bigcup\{A_\alpha : \alpha \in T\} = X$. Then

$$\sum_{x \in X} f(x) = \Sigma\{\sum_{x \in A_\alpha} f(x) : \alpha \in T\}$$

Note that the notation employed in (iv) and (v) is somewhat loose: the same letter f denotes a function on X and also its restriction to a subset of X. However, this notation is commonly used as it is convenient and rarely leads to confusion.

(5-10)[+] Let X be a discrete space, and let $\mathfrak{F} = C_o(X)$ [see (3-1) and (2-6)]. According to (4-4), we can define a Bourbaki integral I on $\mathfrak{F}$ by letting $If = \Sigma_{x \in X} f(x)$ for all $f \in \mathfrak{F}$. Prove

(i) For every function f on X, $\bar{I}^{\#} f = \underline{I}^{\#} f = \Sigma f$ if Σf is defined [see (5-9)], and $\bar{I}^{\#} f = -\underline{I}^{\#} f = +\infty$ otherwise.

(ii) $f \in \mathfrak{L}^{\#}$ iff Σf is defined and it is finite.

(iii) $\bar{I}^* = \bar{I}^{\#}$ and $\underline{I}^* = \underline{I}^{\#}$.

(iv) Let X be infinite, and let $f_n(x) = -1/n$, $n = 1, 2, \ldots$. Then $f_n \in \mathfrak{F}^{\#}_{-}$, $If_n = -\infty$, and $f_n \nearrow 0$.

(5-11) Let I be any of the Bourbaki integrals defined in exercises (4-6)-(4-10). Show that $\bar{I}^* = \bar{I}^{\#}$ and $\underline{I}^* = \underline{I}^{\#}$.

(5-12) Let $X = R$, $\mathfrak{F}_1 = C_o(X)$, and let $\mathfrak{F}_2$ consist of all finite continuous functions on X vanishing exponentially at infinity [see (2-9) (vii)]. According to (4-8), we can define a Bourbaki integral I_2 on $\mathfrak{F}_2$ by letting

$$I_2 f = \lim_{a \to +\infty} \int_{-a}^{a} f(x)\, dx$$

for all $f \in \mathfrak{F}_2$. By I_1 we shall denote the integral I_2 restricted to $\mathfrak{F}_1$. Clearly, I_1 is a Bourbaki integral on $\mathfrak{F}_1$. Show that $\bar{I}_1 = \bar{I}_2$ and $\underline{I}_1 = \underline{I}_2$.

<u>(5-13)</u>[*+] Let $a, b \in \mathbf{R}$, $a < b$, $X = [a,b]$, $\mathfrak{F} = C(X)$, and let g be a non-decreasing real-valued function on X. According to (4-6), we can define a Bourbaki integral on $\mathfrak{F}$ by letting

$$If = \int_a^b f(x)\, dg(x)$$

for all $f \in \mathfrak{F}$.

(i) Prove that every bounded Baire function on X [see (3-13)] belongs to $\mathfrak{L}$.

(ii) Denote by Q the set of all rational points from (a,b), and let $f(x) = 1$ if $x \in Q$ and $f(x) = 0$ if $x \in X - Q$. Show that $f \in \mathfrak{L}$ and that

$$If = \sum_{x \in Q} [g(x+) - g(x-)]$$

where $g(x+) = \lim_{t \to x+} g(t)$ and $g(x-) = \lim_{t \to x-} g(t)$. <u>Hint</u>: Use Examples (0.1) and (0.2).

(iii) Let f be a function on X for which $\int_a^b f(x)\, dg(x)$ exists. Show that if g is continuous on X, then $f \in \mathfrak{L}$ and $If = \int_a^b f(x)\, dg(x)$. <u>Hint</u>: If g is continuous on X show that

$$\underline{\int}_a^b h(x)\, dg(x) \le \underline{I}h \le \bar{I}h \le \overline{\int}_a^b h(x)\, dg(x)$$

for each bounded function h on X.

(iv) Choose $c \in (a,b)$, and let $g(x) = 0$ if $x \in [a,c)$, $g(x) = 1$ if $x \in (c,b]$, and $g(c) = 1/2$. Moreover, let $f(x) = 0$ if $x \in [a,c)$, and $f(x) = 1$ if $x \in [c,b]$. Show that $\int_a^b f(x)\, dg(x)$ exists, $f \in \mathfrak{L}$, $If = 1$, and $\int_a^b f(x)\, dg(x) = 1/2$.

(v) Let $F(a) = 0$, and let

$$F(x) = (x - a)^2 \sin(x - a)^{-2}$$

for each $x \in \mathbf{R}$, $x \neq a$. Prove that F is everywhere differentiable. Hint: Use the definition of the derivative to calculate $F'(a)$.

(vi) For each $x \in [a,b]$, let $g(x) = x$ and $f(x) = F'(x)$. Show that f is not summable. Hint: Observe that $|f| \in \mathfrak{I}_+$ and $I(|f|) = +\infty$.

6. THE FINAL EXTENSION; INTEGRABLE FUNCTIONS

If we restrict ourselves to functions whose integrals are finite, the system $\mathfrak{L}$ of summable functions is a completely adequate extension of the fundamental system $\mathfrak{F}$ and no further extension is necessary. However, we shall frequently encounter functions with infinite integrals, e.g., some functions from $\bar{\mathfrak{F}}$. Therefore, in this chapter we shall extend $\mathfrak{L}$ by adding to it a large family of functions whose integrals are infinite.

Throughout this chapter we shall assume that X is an arbitrary set, $\mathfrak{F}$ is a fundamental system on X, and that I is an integral on $\mathfrak{F}$ which has been extended to $\mathfrak{L} \cup \bar{\mathfrak{F}}$.

We denote by $\mathfrak{L}_+$ the system of all functions f on X such that $f \wedge g \in \mathfrak{L}$ for all $g \in \mathfrak{F}$. Analogously, by $\mathfrak{L}_-$ we denote the system of all functions f on X such that $f \vee g \in \mathfrak{L}$ for all $g \in \mathfrak{F}$.

<u>(6.1)</u> <u>Definition</u>. A function f on X is called <u>integrable</u> whenever it belongs to $\bar{\mathfrak{L}} = \mathfrak{L}_+ \cup \mathfrak{L}_-$.

When carefully looked at, the definition of $\bar{\mathfrak{L}}$ becomes rather intuitive. In essence, $\mathfrak{L}_+$ consists of those functions on X which would belong to $\mathfrak{L}$ if their positive values were not too large [see (6.4)]. Similarly, $\mathfrak{L}_-$ consists of those functions on X which would belong to $\mathfrak{L}$ if their negative values were not too large.

The following are immediate consequences of the definitions.

<u>(6.2)</u> $$f \in \mathfrak{L}_+ \Leftrightarrow -f \in \mathfrak{L}_-$$

<u>(6.3)</u> $$f, g \in \mathfrak{L}_+ \Rightarrow (f \vee g \in \mathfrak{L}_+ \text{ and } f \wedge g \in \mathfrak{L}_+)$$

<u>(6.4)</u> $$f \in \mathfrak{L}_+ \Rightarrow (f^+ \in \mathfrak{L}_+ \text{ and } f^- \in \mathfrak{L})$$

From (2.7), (2.15), their duals, and (5.5) it follows that

(6.5) $$\mathfrak{F}_+ \subset \mathfrak{L}_+ \quad \text{and} \quad \mathfrak{F}_- \subset \mathfrak{L}_-$$

From (5.21), (6.4) and its dual, and (5.22) we obtain

(6.6) $$\mathfrak{L} = \mathfrak{L}_+ \cap \mathfrak{L}_-$$

The next observation is often quite useful.

(6.7) Let f be a nonnegative function on X, and let $f \wedge g \in \mathfrak{L}$ for every nonnegative $g \in \mathfrak{F}$. Then $f \in \mathfrak{L}_+$.

Indeed, $f \wedge g = (f \wedge g^+) - g^-$ whenever f,g are functions on X and $f \geq 0$.

(6.8) Proposition. Let $\{f_n\} \subset \mathfrak{L}_+$, $g \in \mathfrak{L}_+$, and let $f_n \geq g$, $n = 1, 2, \ldots$. If $f = \lim f_n$, then $f \in \mathfrak{L}_+$.

Proof. If $h \in \mathfrak{F}$, then $f_n \wedge h \in \mathfrak{L}$, $g \wedge h \in \mathfrak{L}$,

$$g \wedge h \leq f_n \wedge h \leq h$$

$n = 1, 2, \ldots$, and $f_n \wedge h \to f \wedge h$. Thus by (5.17), $f \wedge h \in \mathfrak{L}$. ///

(6.9) Proposition. If $f \in \mathfrak{L}_+$, then $f \wedge g \in \mathfrak{L}$ for every $g \in \mathfrak{L}$.

Proof. Let $f \in \mathfrak{L}_+$ and $g \in \mathfrak{L}$. Choose $\gamma \in \mathfrak{F}_+$ such that $\gamma \geq g$ and $I\gamma < +\infty$. We can find $\gamma_n \in \mathfrak{F}$ for which $\gamma_n \leq \gamma$ and

$$I\gamma_n > I\gamma - \frac{1}{n}$$

$n = 1, 2, \ldots$. Letting $h_n = \vee_{i=1}^{n} \gamma_i$, we have $h_n \in \mathfrak{F}$ and

$$\gamma_n \leq h_n \leq h_{n+1} \leq \gamma$$

Thus $h_n \nearrow h$ for some function $h \in \mathfrak{F}_+$, $h \leq \gamma$, and $Ih = I\gamma$. Since $f \wedge h_n \in \mathfrak{L}$, $n = 1, 2, \ldots$, $f \wedge h_n \nearrow f \wedge h$, and $f \wedge h \leq \gamma$, it follows from (5.17) that $f \wedge h \in \mathfrak{L}$. For $x \in X$, let

$$\varphi(x) = \gamma(x) - h(x) \quad \text{and} \quad \psi(x) = f(x) \wedge \gamma(x) - f(x) \wedge h(x)$$

whenever the right sides have meaning, and $\varphi(x) = \psi(x) = 0$ otherwise. By (5.14) and (5.15), $\varphi \in \mathfrak{L}$ and

$$I\varphi = I\gamma - Ih = 0$$

Distinguishing the cases $f(x) \geq \gamma(x)$, $f(x) \leq h(x)$, and $h(x) < f(x) < \gamma(x)$, we can see that $0 \leq \psi \leq \varphi$ [see (1-6)(iii)]. Hence,

$$0 \leq \underline{I}\psi \leq \overline{I}\psi \leq I\varphi = 0$$

and $\psi \in \mathfrak{L}$. Since $h \leq \gamma$, $\psi(x) + f(x) \wedge h(x)$ has meaning for all $x \in X$ and $f \wedge \gamma = \psi + f \wedge h$. Thus $f \wedge \gamma \in \mathfrak{L}$. Because

$$f \wedge g = f \wedge (\gamma \wedge g) = (f \wedge \gamma) \wedge g$$

also $f \wedge g \in \mathfrak{L}$. ///

Notice that for a Daniell integral only, the previous proof can be considerably simplified by choosing $\gamma_n \in \mathfrak{F}$ so that $\gamma_n \nearrow \gamma$.

According to (6.6), it is consistent to define a map

$$I : \overline{\mathfrak{L}} \to \mathbf{R}^-$$

by letting $If = \overline{I}f$ if $f \in \mathfrak{L}_+$ and $If = \underline{I}f$ if $f \in \mathfrak{L}_-$.

It follows from (6.5) that this is so far the largest extension of the original integral I on $\mathfrak{F}$. Moreover, in certain sense, it is the largest possible extension. We shall explain this in detail.

From (6.5) and (5.4) we obtain the following.

(6.10) If f is a function on X, then

$$\overline{I}f = \inf\{Ig : g \in \mathfrak{L}_+ , g \geq f\}$$

and

$$\underline{I}f = \sup\{Ig : g \in \mathfrak{L}_- , g \leq f\}$$

Therefore, we cannot define any new upper integral by using $\mathfrak{L}_+$ instead of $\mathfrak{F}_+$, and neither can we define any new lower integral by using $\mathfrak{L}_-$ instead of $\mathfrak{F}_-$. This implies that the extension process we have defined is closed.

It is important to keep in mind that we have defined two, in general different, extensions of an integral I on $\mathfrak{F}$ according to whether I is a Daniell or a Bourbaki integral. The difference occurs in the first step: If I is a Daniell integral, it is extended to $\bar{\mathfrak{F}}^*$; if I is a Bourbaki integral, it is extended to $\bar{\mathfrak{F}}^{\#}$. Although the subsequent extensions are formally identical for both integrals, the initial difference makes it occasionally necessary to distinguish at all levels. For this reason we shall introduce the following self-explanatory symbols: $\bar{I}^*$, $\bar{I}^{\#}$; $\underline{I}^*$, $\underline{I}^{\#}$; $\mathfrak{L}^*$, $\mathfrak{L}^{\#}$; $\mathfrak{L}_+^*$, $\mathfrak{L}_+^{\#}$; $\mathfrak{L}_-^*$, $\mathfrak{L}_-^{\#}$; $\bar{\mathfrak{L}}^*$, $\bar{\mathfrak{L}}^{\#}$. Throughout the book we shall use these symbols freely without any further explanations.

We shall prove a few elementary properties of I on $\bar{\mathfrak{L}}$.

<u>(6.11)</u> $$f \in \mathfrak{L}_+ \Rightarrow \underline{I}f > -\infty$$

<u>Proof</u>. Since $f \geq -f^-$, it suffices to apply (6.4). ///

<u>(6.12)</u> $$(f \in \mathfrak{L}_+ \text{ and } If < +\infty) \Rightarrow f \in \mathfrak{L}$$

<u>Proof</u>. Let $f \in \mathfrak{L}_+$ and $If < +\infty$. Then there is a $g \in \mathfrak{F}_+$ such that $g \geq f$ and $Ig < +\infty$. By (5.5), $g \in \mathfrak{L}$, and it follows from (6.9) that $f = f \wedge g$ belongs to $\mathfrak{L}$. ///

<u>(6.13)</u> $$(f, g \in \bar{\mathfrak{L}} \text{ and } f \leq g) \Rightarrow If \leq Ig$$

<u>Proof</u>. The statement reduces to (5.4) if $f, g \in \mathfrak{L}_+$. If $f \in \mathfrak{L}_-$, then

$$If = \underline{I}f \leq \underline{I}g \leq \bar{I}g$$

for each $g \geq f$, and the statement is correct again. Hence, let $f \in \mathfrak{L}_+$, $g \in \mathfrak{L}_-$, and $f \leq g$. By (6.11) and its dual,

$$Ig = \underline{I}g \geq \underline{I}f > -\infty \qquad \text{and} \qquad If = \bar{I}f \leq \bar{I}g < +\infty$$

Thus by (6.12) and its dual, $f, g \in \mathfrak{L}$, and we have $If \leq Ig$. ///

<u>(6.14)</u> If $f \in \mathfrak{L}_+$ and $0 \leq a < +\infty$, then $af \in \mathfrak{L}_+$. In general, if $f \in \bar{\mathfrak{L}}$ and $a \in \mathbf{R}$, then $af \in \bar{\mathfrak{L}}$ and $I(af) = aIf$.

<u>Proof</u>. Let $f \in \mathfrak{L}_+$ and $0 \leq a < +\infty$. If $g \in \mathfrak{F}$ and $a > 0$, then $g/a \in \mathfrak{F}$. Hence

$$(af) \wedge g = a(f \wedge \frac{g}{a})$$

belongs to $\mathfrak{L}$ and so $af \in \mathfrak{L}_+$. If $a = 0$ then, clearly, $af \in \mathfrak{L}_+$; for $af = 0$.

The general case follows immediately from the first part of the proof, (6.2), (5.2), and (5.3). ///

The proof of the next proposition is based on the following observations [see (1-6)(i and ii)]:

(i) If a, b and c belong to $[0, +\infty]$, then

$$(a + b) \wedge c = (a \wedge c + b \wedge c) \wedge c$$

(ii) If a, b and c belong to $(-\infty, +\infty]$, then $a - b$ has meaning iff $a \wedge (b + c) - b$ has, and

$$(a - b) \wedge c = a \wedge (b + c) - b$$

A straightforward verification will prove both equalities.

(6.15) Proposition. Let $f, g \in \mathcal{L}_+$, and let h be a function on X. If $h(x) = f(x) + g(x)$ for each $x \in X$ for which $f(x) + g(x)$ has meaning, then $h \in \mathcal{L}_+$ and

$$Ih = If + Ig$$

Proof. Letting $\varphi = f^+ + g^+$ and $\psi = f^- + g^-$, we have

$$h(x) = \varphi(x) - \psi(x)$$

for each $x \in X$ for which $\varphi(x) - \psi(x)$ has meaning. By (6.4) and (5.15), $\psi \in \mathcal{L}$. Let $u \in \mathfrak{F}$ and $u \geq 0$. Using (i), we obtain

$$\varphi \wedge u = (f^+ \wedge u + g^+ \wedge u) \wedge u$$

Since the right side belongs to $\mathcal{L}$, it follows from (6.7) that $\varphi \in \mathcal{L}_+$. Given $v \in \mathfrak{F}$, we have

$$h(x) \wedge v(x) = \varphi(x) \wedge [\psi(x) + v(x)] - \psi(x)$$

for each $x \in X$ for which the right side has meaning [see (ii)]. Since $\varphi \in \mathcal{L}_+$ and $\psi \in \mathcal{L}$, it follows from (6.9) that $h \wedge v \in \mathcal{L}$. Therefore $h \in \mathcal{L}_+$, and from (5.6) we obtain

$$Ih \leq If + Ig$$

If $Ih = +\infty$, then $Ih = If + Ig$. Hence, assume $Ih < +\infty$, which implies $h \in \mathcal{L}$ [see (6.12)]. Because $\varphi(x) = h(x) + \psi(x)$ for each $x \in X$ for which $h(x) + \psi(x)$ has meaning, by (5.15), $\varphi \in \mathfrak{L}$. Since $f^+ \leq \varphi$ and $g^+ \leq \varphi$, (6.12) implies $f^+, g^+ \in \mathfrak{L}$. By (6.4) and (5.22) we have $f, g \in \mathfrak{L}$, and thus again $Ih = If + Ig$. ///

(6.16) Corollary. If $f \in \bar{\mathfrak{L}}$, then $|f| \in \mathfrak{L}_+$ and $|If| \leq I(|f|)$.

Proof. By (6.4), its dual, (6.14), and (6.15), we have $|f| \in \mathfrak{L}_+$ and

$$|If| = |If^+ - If^-| \leq If^+ + If^- = I(|f|) \quad ///$$

(6.17) Proposition. Let $f_n \in \mathfrak{L}_+$, $n = 1, 2, \ldots$, and let $f_n \nearrow f$ for some function f on X. Then $f \in \mathfrak{L}_+$ and $If_n \nearrow If$.

This proposition follows immediately from (6.8) and (5.9).

(6.18) Corollary. Let $f_n \in \mathfrak{L}_+$ and $f_n \geq 0$, $n = 1, 2, \ldots$. If $f = \Sigma_{n=1}^{\infty} f_n$, then $f \in \mathfrak{L}_+$ and

$$If = \Sigma_{n=1}^{\infty} If_n$$

Proof. By (6.15), letting $g_n = \Sigma_{i=1}^{n} f_i$, we have $g_n \in \mathfrak{L}_+$ and

$$Ig_n = \Sigma_{i=1}^{n} If_i$$

Since $g_n \nearrow f$, it suffices to apply (6.17). ///

The next proposition is often useful for understanding the extent of the system $\bar{\mathfrak{L}}$.

(6.19) Proposition. Let the function identically equal to $+\infty$ on X belong to $\mathfrak{F}_+^*$. Then a function f on X belongs to $\mathfrak{L}_+$ or $\mathfrak{L}_-$ iff there is a sequence $\{f_n\} \subset \mathfrak{L}$ such that $f_n \nearrow f$ or $f_n \searrow f$, respectively.

Proof. Let $\{g_n\} \subset \mathfrak{F}$ and $g_n \nearrow +\infty$. If $f \in \mathfrak{L}_+$ or $f \in \mathfrak{L}_-$, then, respectively, $f \wedge g_n$ or $f \vee (-g_n)$ belongs to $\mathfrak{L}$, $n = 1, 2, \ldots$. Clearly, $f \wedge g_n \nearrow f$ and $f \vee (-g_n) \searrow f$.

The converse follows from (6.17) and its dual. ///

An important case when the assumption of Proposition (6.19) is always satisfied will be brought up in (6-12)(iii).

If $f \in \mathcal{L}_+ - \mathcal{L}$, then in general, $\underline{I}f < \bar{I}f$ [see (6-1)(iii) and (6-9)(iii)]. However, there is an important special case in which this cannot happen.

(6.20) Proposition. Let $f \in \bar{\mathcal{L}}$, and let there be a monotone sequence $\{f_n\} \subset \mathcal{L}$ such that $f_n \to f$. Then $\underline{I}f = \bar{I}f$.

Proof. Let $f \in \mathcal{L}_+$. If $f_n \searrow f$ then by (6.12), $f \in \mathcal{L}$ and the proposition holds. Thus assume that $f_n \nearrow f$. Using (5.4), (5.2), and (5.9), we obtain

$$\lim If_n \leq \underline{I}f \leq \bar{I}f = \lim If_n$$

For $f \in \mathcal{L}_-$, the proposition follows from the first part of the proof, (6.2), and (5.2). ///

We shall close this section with some observations on how different integrals can be compared.

Suppose that $\mathfrak{F}_1$ and $\mathfrak{F}_2$ are fundamental systems on X and that I_1 and I_2 are integrals on $\mathfrak{F}_1$ and $\mathfrak{F}_2$, respectively. We shall assume that if I_1 is a Bourbaki integral, so is I_2; if I_1 is a Daniell integral, I_2 can be either a Daniell or a Bourbaki integral. The symbols $\mathfrak{F}_{i+}$, $\mathfrak{F}_{i-}$, $\bar{\mathfrak{F}}_i$, $\bar{I}_i$, $\underline{I}_i$, $\mathcal{L}_i$, $\mathcal{L}_{i+}$, $\mathcal{L}_{i-}$, and $\bar{\mathcal{L}}_i$, $(i = 1,2)$ will have the obvious meanings.

(6.21) Lemma. Let $\mathfrak{F}_1 \subset \mathfrak{F}_2$, and let $I_1 f = I_2 f$ for each $f \in \mathfrak{F}_1$. Then

$$\underline{I}_1 f \leq \underline{I}_2 f \leq \bar{I}_2 f \leq \bar{I}_1 f$$

for every function f on X.

Proof. Clearly, $\mathfrak{F}_{1+} \subset \mathfrak{F}_{2+}$ and $\mathfrak{F}_{1-} \subset \mathfrak{F}_{2-}$. By (4.16) and (4.17), $I_1 f = I_2 f$ for each $f \in \bar{\mathfrak{F}}_1$. The inequality follows. ///

(6.22) Corollary. Let $\mathfrak{F}_1 \subset \mathfrak{F}_2$, and let $I_1 f = I_2 f$ for each $f \in \mathfrak{F}_1$. Then $\mathcal{L}_1 \subset \mathcal{L}_2$ and $I_1 f = I_2 f$ for each $f \in \mathcal{L}_1$.

Let I be a Bourbaki integral on $\mathfrak{F}$. Then by (4.6), I is also a Daniell integral on $\mathfrak{F}$. Following our notation, we shall denote by I^* the integral I viewed as a Daniell integral, and by $I^{\#}$ the integral I viewed as a Bourbaki integral. In (6.22), letting $\mathfrak{F}_1 = \mathfrak{F}_2 = \mathfrak{F}$, $I_1 = I^*$, and $I_2 = I^{\#}$, we obtain the following corollary.

<u>(6.23)</u> <u>Corollary</u>. Let I be a Bourbaki integral on $\mathfrak{F}$. Then $\mathfrak{L}^* \subset \mathfrak{L}^{\#}$, $\mathfrak{L}^*_+ \subset \mathfrak{L}^{\#}_+$, $\mathfrak{L}^*_- \subset \mathfrak{L}^{\#}_-$, and $I^*f = I^{\#}f$ for each $f \in \mathfrak{L}^*$.

<u>(6.24)</u> <u>Remark</u>. Without some additional hypotheses [see, e.g., (6-6)-(6-8)], the assumptions of Corollary (6.22) do not imply that $\bar{\mathfrak{L}}_1 \subset \bar{\mathfrak{L}}_2$ [see (6-3)(i)]. Also if I is a Bourbaki integral on $\mathfrak{F}$, it does not follow that $I^*f = I^{\#}f$ for each $f \in \bar{\mathfrak{L}}^*$ [see (6-1)(iii)].

An obvious goal in extending integrals is to obtain as large an extension as possible. With this goal in mind, two important practical conclusions can be drawn from Corollaries (6.22) and (6.23):

(i) We should begin with as large a fundamental system as possible

(ii) Bourbaki integrals should be extended as Bourbaki integrals.

For an illustration of these rules, we refer the reader to exercises (6-1), (6-5), and (6-9).

In accordance with rule (ii), unless explicitly stated otherwise, we shall never extend a Bourbaki integral as a Daniell integral. Thus if I is a Bourbaki integral on a fundamental system $\mathfrak{F}$, the symbols $\mathfrak{F}_+$, $\mathfrak{F}_-$, $\bar{\mathfrak{F}}$, $\underline{I}$, $\bar{I}$, $\mathfrak{L}$, $\mathfrak{L}_+$, $\mathfrak{L}_-$, and $\bar{\mathfrak{L}}$ will always mean $\mathfrak{F}^{\#}_+$, $\mathfrak{F}^{\#}_-$, $\bar{\mathfrak{F}}^{\#}$, $\underline{I}^{\#}$, $\bar{I}^{\#}$, $\mathfrak{L}^{\#}$, $\mathfrak{L}^{\#}_+$, $\mathfrak{L}^{\#}_-$, and $\bar{\mathfrak{L}}^{\#}$, respectively.

<u>Exercises</u>

In exercises (6-1) and (6-2), let X be a discrete space, and let $\mathfrak{F} = C_o(X)$.

<u>(6-1)</u> Choose $x_o \in X$, and let I be the Dirac integral at x_o [see (4-1)]. Use (5-4) to show that

(i) $\mathcal{L}_+^*$ consists of all functions f on X for which $f(x_o) > -\infty$ and the set $\{x \in X : f(x) < 0\}$ is countable.

(ii) $\mathcal{L}_+^\#$ consist of all functions f on X for which $f(x_o) > -\infty$.

(iii) If $f \in \mathcal{L}_+^* - \mathcal{L}^*$, then $I^*f = +\infty$ and $\underline{I}^*f = \overline{I}^\# = f(x_o)$.

(6-2)$^+$ For $f \in \mathfrak{F}$, let $If = \Sigma_{x \in X} f(x)$ [see (4-4)]. Using (5-10), show that

(i) $f \in \overline{\mathfrak{L}}$ iff Σf is defined [see (5-9)].

(ii) $If = \Sigma f$ for each $f \in \overline{\mathfrak{L}}$.

(6-3) Let $a, b \in \mathbf{R}$, $a < b$, and $X = [a,b]$. Let $\mathfrak{F}_1$ consist only of the zero function on X, and let $\mathfrak{F}_2 = \mathcal{C}(X)$. For $f \in \mathfrak{F}_2$, set

$$I_2 f = \int_a^b f(x)\, dx$$

[see (4-6)], and let I_1 be the restrictions of I_2 to $\mathfrak{F}_1$.

(i) In (10-8) we shall prove the existence of a nonnegative function f on X which does not belong to $\mathcal{L}_{2+}$. Use this fact now to show that $\mathcal{L}_{1+} \not\subset \mathcal{L}_{2+}$.

(ii) For $x \in X$, let $f(x) = 0$ if x is irrational and $f(x) = 1$ otherwise. Show that $f \in \mathcal{L}_{1+} \cap \mathcal{L}_{2+}$, and calculate $I_1 f$ and $I_2 f$. Hint: Use (5-13)(ii).

(6-4)$^+$ Let X be a locally compact Hausdorff space, $\mathfrak{F} = C_o(X)$, and let I be a Bourbaki integral on $\mathfrak{F}$. Show that every nonnegative Baire function on X [see (3-13)] belongs to $\mathcal{L}_+$.

(6-5)* Let $X = [0,1] \times \{0,1\}$ be the topological space from exercise (3-3), and let $\mathfrak{F} = \mathcal{C}(X)$. For $f \in \mathfrak{F}$, set

$$If = \int_0^1 f(s,0)\, ds$$

and show that I is a Bourbaki integral on $\mathfrak{F}$. For each $(s,t) \in X$, let $f(s,t) = t$. Prove

(i) $f \in \mathcal{L}^\#$ and $I^\# f = 0$.

(ii) $\underline{I}^*f = 0$ and $\bar{I}^*f = 1$.

In exercises (6-6) through (6-10) we shall assume that $\mathfrak{F}_1$ and $\mathfrak{F}_2$ are fundamental systems on a set X and that I_1 and I_2 are integrals on $\mathfrak{F}_1$ and $\mathfrak{F}_2$, respectively. Moreover, we shall assume that $I_1f = I_2f$ for each $f \in \mathfrak{F}_1 \cap \mathfrak{F}_2$.

(6-6) Suppose that I_1 is a Daniell integral or I_2 is a Bourbaki integral. Prove the following:

(i) Let $\mathfrak{F}_1 \subset \mathfrak{F}_2 \subset \mathfrak{L}_1$. Then $\bar{\mathfrak{L}}_1 \subset \bar{\mathfrak{L}}_2$ and $I_1f = I_2f$ for each $f \in \mathfrak{L}_1$. Hint: Use (6.22).

(ii) Let $\mathfrak{F}_1 \subset \mathfrak{F}_2$, and let the function identically equal to $+\infty$ on X belong to $\mathfrak{F}^*_{1+}$. Then $\bar{\mathfrak{L}}_1 \subset \bar{\mathfrak{L}}_2$ and $I_1f = I_2f$ for each $f \in \bar{\mathfrak{L}}_1$. Hint: Use (6.19).

(6-7)* Suppose that I_2 is a Daniell integral, $\mathfrak{F}_1 \subset \mathfrak{F}_2 \subset \mathfrak{L}_1$, and that $I_1f = I_2f$ for each $f \in \mathfrak{F}_2$. Prove

(i) For every function f on X,

$$\underline{I}_2 f \le \underline{I}_1 f \le \bar{I}_1 f \le \bar{I}_2 f$$

Hint: Observe that $\bar{\mathfrak{F}}_2 \subset \bar{\mathfrak{L}}_1$ and $I_1f = I_2f$ for each $f \in \bar{\mathfrak{F}}_2$.

(ii) $\mathfrak{L}_2 \subset \mathfrak{L}_1$, $\mathfrak{L}_{2+} \subset \mathfrak{L}_{1+}$, $\mathfrak{L}_{2-} \subset \mathfrak{L}_{1-}$, and $I_1f = I_2f$ for each $f \in \mathfrak{L}_2$.

(iii) Let $\mathfrak{F}_2$ consist of all finite functions from $\mathfrak{L}_1$. Then $\bar{\mathfrak{L}}_1 = \bar{\mathfrak{L}}_2$ and $I_1f = I_2f$ for each $f \in \bar{\mathfrak{L}}_1$. Hint: If $\bar{I}_1 f < +\infty$, use (6.10), (5.16), and the hint from (i) to show that

$$\begin{aligned} \bar{I}_1 f &= \inf\{I_1 g : g \in \mathfrak{L}_1 ,\ -\infty \ne g \ge f\} \\ &\ge \inf\{I_1 g : g \in \mathfrak{F}_{2+} ,\ g \ge f\} = \bar{I}_2 f \end{aligned}$$

Combine this with (i).

Notice that (iii) gives another indication that our extension process is closed. The final, and perhaps the most convincing, arguments to this extent will be given in (11.9) and (11-9)(vi).

Also observe that (ii) provides a new proof to (6.23).

(6-8) Suppose that both I_1 and I_2 are Daniell integrals. Prove the following:

(i) Let $\mathfrak{F}_1 \subset \mathfrak{L}_2$, $\mathfrak{F}_2 \subset \mathfrak{L}_1$, and let $I_1 f = I_2 f$ for each $f \in \mathfrak{F}_1 \cup \mathfrak{F}_2$. Then $\bar{\mathfrak{L}}_1 = \bar{\mathfrak{L}}_2$ and $I_1 f = I_2 f$ for each $f \in \bar{\mathfrak{L}}_1$. Hint: Observe that $\mathfrak{F}_{1+} \subset \mathfrak{L}_{2+}$, $\mathfrak{F}_{2+} \subset \mathfrak{L}_{1+}$, and $I_1 f = I_2 f$ for each $f \in \mathfrak{F}_{1+} \cup \mathfrak{F}_{2+}$.

(ii) Let $\mathfrak{F}_1 \subset \mathfrak{F}_2 \subset \mathfrak{L}_1$. Then $\bar{\mathfrak{L}}_1 = \bar{\mathfrak{L}}_2$ and $I_1 f = I_2 f$ for each $f \in \bar{\mathfrak{L}}_1$. Hint: Use (6-6)(i) to show that $I_1 f = I_2 f$ for each $f \in \mathfrak{F}_2$ and apply (i).

(6-9) Give X the discrete topology, and choose $A \subset X$. Let $\mathfrak{F}_2 = C_o(X)$ and $\mathfrak{F}_1 = \{f \in C_o(X) : \text{supp } f \subset A\}$. For $f \in \mathfrak{F}_2$, set $I_2 f = \Sigma_{x \in X} f(x)$ [see (4-4)], and let I_1 be the restriction of I_2 to $\mathfrak{F}_1$. Show that

(i) $\mathfrak{L}_1 = \{f \in \mathfrak{L}_2 : \text{supp } f \subset A\}$.

(ii) $\mathfrak{L}_{1+} = \{f \in \mathfrak{L}_{2+} : \text{supp } f^- \subset A\}$.

(iii) If $f \in \mathfrak{L}_{1+} \cap \mathfrak{L}_2 - \mathfrak{L}_1$, then $I_1 f = +\infty$ and $\underline{I}_1 f \le I_2 f < +\infty$.

(6-10) Let $X = \mathbf{R}$, $\mathfrak{F}_1 = C_o(X)$, and let $\mathfrak{F}_2$ consist of all finite continuous functions on X vanishing exponentially at infinity [see (2-9)(vii)]. For $f \in \mathfrak{F}_2$, set

$$I_2 f = \lim_{a \to +\infty} \int_{-a}^{a} f(x)\, dx$$

[see (4-8)], and let I_1 be the restriction of I_2 to $\mathfrak{F}_1$. Show that $\bar{\mathfrak{L}}_1 = \bar{\mathfrak{L}}_2$ and $I_1 f = I_2 f$ for each $f \in \bar{\mathfrak{L}}_1$. Hint: Observe that $\mathfrak{F}^*_{i+} = \mathfrak{F}^\#_{i+}$ $(i = 1,2)$, and apply (6-8)(ii).

In exercises (6-11) and (6-12) we shall denote by φ_1 and φ_∞ the functions on X which are identically equal to 1 and $+\infty$, respectively.

(6-11) Let $\mathfrak{F}$ be a fundamental system on a set X.

(i) Show that $\varphi_\infty \in \mathfrak{F}_+$ whenever $\varphi_1 \in \mathfrak{F}_+$.

(ii) Consider the system $\mathfrak{F}$ from (2-8)(b) restricted to $X = (0,1]$, and observe that the converse of (i) is generally false.

(6-12) Let X be a locally compact Hausdorff space, and let $\mathfrak{F} = C_o(X)$. Show that

(i) φ_1, $\varphi_\infty \in \mathfrak{F}_+^\#$.

(ii) $\varphi_1 \in \mathfrak{F}_+^* \iff \varphi_\infty \in \mathfrak{F}_+^*$.

(iii) $\varphi_\infty \in \mathfrak{F}_+^*$ iff X is σ-compact.

(6-13) Let $\{f_n\} \subset \mathfrak{L}_+$. Use (6.3) and (6.8) to show that

(i) $\sup f_n \in \mathfrak{L}_+$.

(ii) If $f_n \geq g$, $n = 1, 2, \ldots$, for some $g \in \mathfrak{L}_+$, then $\inf f_n$, $\limsup f_n$, and $\liminf f_n$ belong to $\mathfrak{L}_+$.

(6-14)$^+$ Let $\{f_\alpha : \alpha \in T\}$ be a family of nonnegative integrable functions on X, and let $f = \Sigma_{\alpha \in T} f_\alpha$.

(i) Prove that

$$\sum_{\alpha \in T} If_\alpha \leq \bar{I}f$$

Hint: Use the definition of $\Sigma_{\alpha \in T} f_\alpha$ [see (5-9)] and (6.15).

(ii) Show by example that the inequality in (i) may be sharp even if f is a summable function.

7. MEASURABLE FUNCTIONS

In the previous chapters we successfully extended the fundamental system $\mathfrak{F}$ into a large family $\bar{\mathfrak{L}}$ of integrable functions. Defining $\bar{\mathfrak{L}}$ was, however, a lengthy and complicated process which was constructive at its first stage (the definition of $\bar{\mathfrak{F}}$) and descriptive otherwise. Consequently, the structure of $\bar{\mathfrak{L}}$ remains rather obscure at this point. In particular, we have no convenient means how to decide whether a given function f on X belongs to $\bar{\mathfrak{L}}$. We shall try to remedy this situation in the following chapters. Our main goal will be to describe the functions in $\bar{\mathfrak{L}}$ in terms of a certain family of sets. This is a common approach based on the idea that sets may be easier to understand than functions [see (7.20)].

<u>(7.1)</u> <u>Definition</u>. A family $\mathfrak{M}$ of subsets of a set X is called a σ-<u>algebra</u> in X if the following conditions hold:

(i) The set X belongs to $\mathfrak{M}$

(ii) If $A \in \mathfrak{M}$, then $X - A \in \mathfrak{M}$

(iii) If $\{A_n\} \subset \mathfrak{M}$ is a countable family, then $\bigcup_n A_n \in \mathfrak{M}$.

Let $\mathfrak{M}$ be a σ-algebra in a set X. It follows from (i) and (ii) that the empty set $\emptyset = X - X$ belongs to $\mathfrak{M}$. If $\{A_n\} \subset \mathfrak{M}$ is a countable family, then the intersection

$$\bigcap_n A_n = X - \bigcup_n (X - A_n)$$

belongs to $\mathfrak{M}$. Thus if $A, B \in \mathfrak{M}$, then $A - B = A \cap (X - B)$ belongs to $\mathfrak{M}$.

The families exp X (see Chapter 1, Section A) and $\{\emptyset, X\}$ are the extreme examples of σ-algebras in X.

The following proposition is an immediate consequence of Definition (7.1).

(7.2) Proposition. Let $\{\mathfrak{M}_\alpha : \alpha \in T\}$ be a nonempty family of σ-algebras in a set X. Then $\bigcap\{\mathfrak{M}_\alpha : \alpha \in T\}$ is also a σ-algebra in X.

(7.3) Corollary. Let X be a set, and let $\mathfrak{C} \subset \exp X$. Then there is a unique σ-algebra $\mathfrak{M}$ in X such that $\mathfrak{C} \subset \mathfrak{M}$ and if $\mathfrak{M}'$ is a σ-algebra in X containing $\mathfrak{C}$, then $\mathfrak{M} \subset \mathfrak{M}'$.

Proof. The uniqueness of $\mathfrak{M}$ is obvious. To prove the existence, consider the family $\{\mathfrak{M}_\alpha : \alpha \in T\}$ of all σ-algebras in X containing $\mathfrak{C}$. This family is nonempty, as it contains the σ-algebra exp X, and so it suffices to let $\mathfrak{M} = \bigcap_{\alpha \in T} \mathfrak{M}_\alpha$. ///

The σ-algebra $\mathfrak{M}$ from the previous corollary is called the σ-algebra in X generated by $\mathfrak{C}$. The corollary then states that the σ-algebra in X generated by $\mathfrak{C}$ is the smallest σ-algebra in X containing $\mathfrak{C}$.

(7.4) Definition. Let X be a topological space. The σ-algebra in X generated by the family of all open subsets of X is called the Borel σ-algebra in X, denoted by $\mathfrak{B}$. The elements of $\mathfrak{B}$ are called Borel sets.

It is immediate that the Borel σ-algebra in X is also generated by the family of all closed subsets of X. Sometimes, however, we can find rather small families of subsets of X which still generate the Borel σ-algebra in X. We shall give a useful example.

Let

$$\mathfrak{H}_+ = \{[a,+\infty] : a \in \mathbf{R}\} \qquad \mathfrak{H}_- = \{[-\infty,a] : a \in \mathbf{R}\}$$

$$\mathfrak{H}_+^\circ = \{(a,+\infty] : a \in \mathbf{R}\} \qquad \mathfrak{H}_-^\circ = \{[-\infty,a) : a \in \mathbf{R}\}$$

(7.5) Proposition. Each of the families $\mathfrak{H}_+$, $\mathfrak{H}_-$, $\mathfrak{H}_+^\circ$, and $\mathfrak{H}_-^\circ$ generates the Borel σ-algebra $\mathfrak{B}$ in $\mathbf{R}^-$.

Proof. Because for $a \in \mathbf{R}$,

$$(a,+\infty] = \bigcup_{n=1}^{\infty} [a+\frac{1}{n},+\infty]$$

$$[a,+\infty] = \bigcap_{n=1}^{\infty} (a-\frac{1}{n},+\infty]$$

$$[-\infty,a) = \mathbf{R}^- - [a,+\infty]$$

$$[-\infty,a] = \mathbf{R}^- - (a,+\infty]$$

each of the families $\mathfrak{H}_+$, $\mathfrak{H}_-$, $\mathfrak{H}_+^o$, and $\mathfrak{H}_-^o$ generates the same σ-algebra $\mathfrak{M}$ in $\mathbf{R}^-$. Since every set from $\mathfrak{H}_+$ and $\mathfrak{H}_-$ is closed in $\mathbf{R}^-$ and every set from $\mathfrak{H}_+^o$ and $\mathfrak{H}_-^o$ is open in $\mathbf{R}^-$, we have $\mathfrak{M} \subset \mathfrak{B}$. Finally, every open subset of $\mathbf{R}^-$ is a countable union of open intervals, and every open interval has a form

$$(a,+\infty] \cap [-\infty,b)$$

where $a,b \in \mathbf{R}^-$. Thus $\mathfrak{M}$ contains all open subsets of $\mathbf{R}^-$, and so $\mathfrak{M} = \mathfrak{B}$. ///

Let X and Y be sets, and let $f : X \to Y$ be a map. For $B \subset Y$, let

$$f^{-1}(B) = \{x \in X : f(x) \in B\}$$

If $\mathfrak{N}$ is a family of subsets of Y, then

$$f^{-1}(\mathfrak{N}) = \{f^{-1}(B) : B \in \mathfrak{N}\}$$

is a family of subsets of X. A direct check reveals that if $\mathfrak{N}$ is a σ-algebra in Y, then $f^{-1}(\mathfrak{N})$ is a σ-algebra in X.

We are naturally led to the following definitions.

(7.6) Definition. A measurable space is a pair $(X,\mathfrak{M})$, where X is a set and $\mathfrak{M}$ is a σ-algebra in X. The elements of $\mathfrak{M}$ are called measurable sets.

(7.7) Definition. Let $(X,\mathfrak{M})$ and $(Y,\mathfrak{N})$ be measurable spaces. A map $f : X \to Y$ is called measurable if $f^{-1}(\mathfrak{N}) \subset \mathfrak{M}$.

(7.8) Definition. Let $(X,\mathfrak{M})$ be a measurable space. A function f on X is called measurable if the map $f : X \to \mathbf{R}^-$ is measurable with respect to the Borel σ-algebra $\mathfrak{B}$ in $\mathbf{R}^-$, i.e., if $f^{-1}(\mathfrak{B}) \subset \mathfrak{M}$.

If there is a need to be more specific, we shall say $(\mathfrak{M},\mathfrak{N})$-measurable map instead of measurable map only. Similarly, instead of measurable set and measurable function we shall say $\mathfrak{M}$-measurable set and $\mathfrak{M}$-measurable function, respectively.

The usage of the word "measurable" will be justified later. We shall see that the measurable sets in X are those subsets of X to which, in some sense, a "measure" can be associated. A precise meaning will be given to this rather vague statement in Chapter 8.

The reader familiar with point-set topology will notice the similarity in the definitions of topological and measurable spaces and continuous and measurable maps.

The remainder of this chapter will be devoted to studying measurable functions on a given measurable space. Thus throughout we shall assume that $(X,\mathfrak{M})$ is a measurable space, and we shall denote by $\mathfrak{m}$ the family of all measurable functions on X.

(7.9) Proposition. If f is a function on X, then the following properties are equivalent:

(i) $f \in \mathfrak{m}$
(ii) $\{x \in X : f(x) \geq a\} \in \mathfrak{M}$ for all $a \in \mathbf{R}$
(iii) $\{x \in X : f(x) \leq a\} \in \mathfrak{M}$ for all $a \in \mathbf{R}$
(iv) $\{x \in X : f(x) > a\} \in \mathfrak{M}$ for all $a \in \mathbf{R}$
(v) $\{x \in X : f(x) < a\} \in \mathfrak{M}$ for all $a \in \mathbf{R}$

Proof. It is clear that property (i) implies all others.

Let $\mathfrak{H} \subset \exp \mathbf{R}^-$ generate $\mathfrak{B}$, and let $\mathfrak{N}$ be a σ-algebra in X containing $f^{-1}(\mathfrak{H})$. Then

$$f(\mathfrak{N}) = \{A \subset \mathbf{R}^- : f^{-1}(A) \in \mathfrak{N}\}$$

is a σ-algebra in $\mathbf{R}^-$ containing $\mathfrak{H}$. Thus $\mathfrak{B} \subset f(\mathfrak{N})$, and consequently, $f^{-1}(\mathfrak{B}) \subset \mathfrak{N}$. Since $f^{-1}(\mathfrak{H}) \subset f^{-1}(\mathfrak{B})$, $f^{-1}(\mathfrak{B})$ is the σ-algebra in X generated by $f^{-1}(\mathfrak{H})$. It follows from (7.5) that each of the properties (ii)-(v) implies (i). ///

The previous proposition is very important as it gives a simple way of deciding whether a given function f on X is measurable. It will be used repeatedly in proving other properties of measurable functions.

(7.10) Proposition. Every constant function on X is measurable.

This proposition is an immediate consequence of Definition (7.8). With a bit of work we shall obtain a stronger result.

(7.11) Proposition. Let $f \in \mathfrak{m}$, $A \in \mathfrak{M}$, and $a \in \mathbf{R}^-$. If g is a function on X such that $g(x) = f(x)$ if $x \in A$ and $g(x) = a$ if $x \in X - A$, then $g \in \mathfrak{m}$.

Proof. Choose $c \in \mathbf{R}$. Then

$$\{x \in X : g(x) > c\} = [\{x \in X : f(x) > c\} \cap A] \cup (X - A)$$

if $c < a$, and

$$\{x \in X : g(x) > c\} = \{x \in X : f(x) > c\} \cap A$$

if $c \geq a$. The proposition follows from (7.9). ///

(7.12) Lemma. Let $f, g \in \mathfrak{m}$ and $a \in \mathbf{R}$. Then the set

$$\{x \in X : f(x) > a - g(x)\}$$

is measurable.

Proof. Let Q be the set of all rational numbers. We have

$$\begin{aligned} &\{x \in X : f(x) > a - g(x)\} \\ &\quad = \bigcup_{r \in Q} [\{x \in X : f(x) > r\} \cap \{x \in X : a - g(x) < r\}] \\ &\quad = \bigcup_{r \in Q} [\{x \in X : f(x) > r\} \cap \{x \in X : g(x) > a - r\}] \end{aligned}$$

and the lemma follows form (7.9) and the countability of Q. ///

(7.13) Proposition. Let $f, g \in \mathfrak{m}$ and $a \in \mathbf{R}^-$. Let h be a function on X such that

$$h(x) = f(x) + g(x)$$

for each $x \in X$ for which $f(x) + g(x)$ has meaning, and $h(x) = a$ otherwise. Then $h \in \mathfrak{m}$.

Proof. It follows directly from Definition (7.8) that if $k \in \mathfrak{m}$, then the set

$$\{x \in A : k(x) = a\}$$

is measurable for every $a \in \mathbf{R}^-$. This implies that the set A of all $x \in X$ for which $f(x) + g(x)$ has no meaning is measurable. For $x \in A$, set

$$f_1(x) = a \qquad g_1(x) = 0$$

and for $x \in X - A$, set

$$f_1(x) = f(x) \qquad g_1(x) = g(x)$$

Clearly, $h = f_1 + g_1$, and by (7.11), f_1 and g_1 are measurable. For $c \in \mathbf{R}$, we have

$$\{x \in X : h(x) > c\} = \{x \in X : f_1(x) > c - g_1(x)\}$$

and the proposition follows from (7.12) and (7.9). ///

(7.14) Proposition. If $f \in \mathfrak{m}$ and $a \in \mathbf{R}$, then $af \in \mathfrak{m}$.

Proof. Choose $c \in \mathbf{R}$. Then

$$\{x \in X : af(x) > c\} = \{x \in X : f(x) > \tfrac{c}{a}\}$$

if $a > 0$, and

$$\{x \in X : af(x) > c\} = \{x \in X : f(x) < \tfrac{c}{a}\}$$

if $a < 0$. The proposition follows from (7.9) and (7.10). ///

(7.15) Proposition. If $\{f_n\}$ is a sequence of measurable functions on X, then the functions $\sup f_n$, $\inf f_n$, $\limsup f_n$, are $\liminf f_n$ are also measurable. In particular, if $\lim f_n$ exists, it is measurable.

Proof. For every $a \in \mathbf{R}$, we have

$$\{x \in X : \sup f_n(x) > a\} = \bigcup_{n=1}^{\infty} \{x \in X : f_n(x) > a\}$$

$$\{x \in X : \inf f_n(x) < a\} = \bigcup_{n=1}^{\infty} \{x \in X : f_n(x) < a\}$$

Thus by (7.9), $\sup f_n$ and $\inf f_n$ are measurable. Since

$$\limsup f_n = \inf_k(\sup_{n\geq k} f_n)$$

$$\liminf f_n = \sup_k(\inf_{n\geq k} f_n)$$

the proof is completed. ///

(7.16) Corollary. If f and g are measurable functions on X, then so are $f \vee g$, $f \wedge g$, f^+, f^-, and $|f|$.

Proof. Letting $f_1 = f$ and $f_n = g$ for n = 2, 3, ..., we obtain that $f \vee g$ and $f \wedge g$ are measurable functions on X. To complete the proof it suffices to apply (7.14) and (7.13). ///

(7.17) Proposition. Let $f \in \mathfrak{M}$, $f \geq 0$, and let $a \in (0,+\infty)$. Then $f^a \in \mathfrak{M}$.

Proof. Choose $c \in \mathbf{R}$. Then

$$\{x \in X : f^a(x) \geq c\} = \{x \in X : f(x) \geq c^{1/a}\}$$

if $c > 0$, and $\{x \in X : f^a(x) \geq c\} = X$ if $c \leq 0$. An application of (7.9) completes the proof. ///

(7.18) Proposition. If f and g are measurable functions on X, then so is fg.

Proof. (a) Let $f, g \in \mathfrak{M}$ be positive. Then for $x \in X$,

$$f(x)g(x) = \frac{1}{2}([f(x) + g(x)]^2 - [f^2(x) + g^2(x)])$$

if the right side has meaning, and $f(x)g(x) = +\infty$ otherwise. By (7.13), (7.14), and (7.17), $fg \in \mathfrak{M}$.

(b) Let $f, g \in \mathfrak{M}$ be nonnegative. The set

$$A = \{x \in X : f(x)g(x) = 0\}$$
$$= \{x \in X : f(x) = 0\} \cup \{x \in X : g(x) = 0\}$$

is measurable. Set

$$f_1(x) = f(x) \qquad g_1(x) = g(x)$$

if $x \in X - A$, and $f(x) = g(x) = 1$ if $x \in A$. Using (7.11) and (a), we obtain $fg \in \mathfrak{M}$.

(c) Let $f, g \in \mathfrak{m}$ be arbitrary. Since

$$fg = (f^+ - f^-)(g^+ - g^-) = f^+g^+ + f^-g^- - f^+g^- - f^-g^+$$

(7.16), (7.13), (7.14), and (b) imply that $fg \in \mathfrak{m}$. ///

Measurable functions are, in general, very complicated. In order to understand their structure, we shall try to approximate them by some simpler functions. We used the same idea in Chapter 3 when we approximated lower semicontinuous functions by steplike functions. Up to some technical details the approximation of measurable functions will be virtually identical with that of lower semicontinuous functions.

(7.19) Definition. The characteristic function of a set $A \subset X$ is the function χ_A on X such that

$$\chi_A(x) = \begin{cases} 1 & \text{if } x \in A \\ 0 & \text{if } x \in X - A \end{cases}$$

The following are easy consequences of the definition.

(7.20) The map $A \mapsto \chi_A$ is an imbedding of $\exp X$ into the family of all functions on X.

(7.21) If $A \subset B \subset X$, then $\chi_A \le \chi_B$ and $\chi_{B-A} = \chi_B - \chi_A$.

(7.22) If $\{A_\alpha : \alpha \in T\} \subset \exp X$, $A = \bigcup_{\alpha \in T} A_\alpha$, and $B = \bigcap_{\alpha \in T} A_\alpha$, then $\chi_A = \sup\{\chi_{A_\alpha} : \alpha \in T\}$ and $\chi_B = \inf\{\chi_{A_\alpha} : \alpha \in T\}$.

(7.23) If $\{A_\alpha : \alpha \in T\} \subset \exp X$ is a disjoint family and $A = \bigcup_{\alpha \in T} A_\alpha$, then $\chi_A = \Sigma_{\alpha \in T} \chi_{A_\alpha}$.

Note that in (7.23) for each $x \in X$, $\chi_{A_\alpha}(x) = 0$ for all but at most one $\alpha \in T$. Thus the sum $\Sigma_{\alpha \in T} \chi_{A_\alpha}(x)$ is well defined for each $x \in X$.

(7.24) $A \in \mathfrak{M} \Leftrightarrow \chi_A \in \mathfrak{m}$

(7.25) Definition. A measurable function f on X is called simple whenever the set

$$f(X) = \{f(x) : x \in X\}$$

is a finite subset of $\mathbf{R}$.

The family of all simple functions on X is denoted by $\mathcal{S}$.

Note that the definition of simple function depends on the σ-algebra $\mathfrak{M}$ of measurable sets. If we want to emphasize this, we shall talk about $\mathfrak{M}$-simple functions rather than about simple functions only.

(7.26) Proposition. If $f \in \mathcal{S}$ then there are nonempty, disjoint measurable sets $A_1, \ldots, A_n$ and distinct real numbers $a_1, \ldots, a_n$ such that

$$X = \bigcup_{i=1}^{n} A_i \qquad \text{and} \qquad f = \sum_{i=1}^{n} a_i \chi_{A_i}$$

Up to order, the sets $A_1, \ldots, A_n$ and the numbers $a_1, \ldots, a_n$ are uniquely determined by the function f.

Proof. Let $f(X) = \{a_1, \ldots, a_n\}$ and $A_i = \{x \in X : f(x) = a_i\}$, $i = 1, \ldots, n$. Then $A_1, \ldots, A_n$ are nonempty, disjoint measurable sets whose union is X and $f = \Sigma_{i=1}^{n} a_i \chi_{A_i}$. Let $f = \Sigma_{i=1}^{m} b_i \chi_{B_i}$, where $b_1, \ldots, b_m$ are distinct real numbers and $B_1, \ldots, B_m$ are nonempty, disjoint measurable sets whose union is X. Then

$$f(X) = \{a_1, \ldots, a_n\} = \{b_1, \ldots, b_m\}$$

Since $a_1, \ldots, a_n$ are distinct and so are $b_1, \ldots, b_n$, we have $n = m$. Since $B_i = \{x \in X : f(x) = b_i\}$, $i = 1, \ldots, n$, we also have

$$\{A_1, \ldots, A_n\} = \{B_1, \ldots, B_n\} \quad ///$$

We note that there are usually many different ways to write a simple function as a linear combination of characteristic functions. The previous proposition, however, allows us to single out a specific way.

(7.27) Definition. The canonical form of a simple function f on X is the expression

$$f = \sum_{i=1}^{n} a_i \chi_{A_i}$$

where $a_1, \ldots, a_n$ are distinct real numbers and $A_1, \ldots, A_n$ are non-empty, disjoint measurable sets whose union is X.

The following is a corollary of Proposition (7.26).

(7.28) Corollary. Each simple function has one and only one canonical form.

Next we shall formulate and prove the promised approximation theorem.

(7.29) Theorem. Let $f \in \mathfrak{M}$ and $f \geq 0$. Then there is a sequence $\{f_n\} \subset \mathcal{S}$ such that $f_1 \geq 0$ and $f_n \nearrow f$.

Proof. For $n = 1, 2, \ldots$ and $k = 1, 2, \ldots, n2^n$, let

$$A_n = \{x \in X : f(x) \geq n\}$$

$$A_{n,k} = \{x \in X : (k-1)2^{-n} \leq f(x) < k2^{-n}\}$$

Since these sets are measurable,

$$f_n = n\chi_{A_n} + \sum_{k=1}^{n2^n} (k-1)2^{-n}\chi_{A_{n,k}}$$

$n = 1, 2, \ldots,$ are simple functions. Without difficulty we can check that $0 \leq f_n \leq f_{n+1} \leq f$. Moreover, for $x \in X$,

$$f(x) - f_n(x) \leq 2^{-n}$$

if $f(x) < n$, and $f_n(x) = n$ if $f(x) \geq n$. ///

The idea of the previous proof is illustrated in Fig. 7.1. The reader should compare the proof of Theorem (7.29) with that of Lemma (3.9).

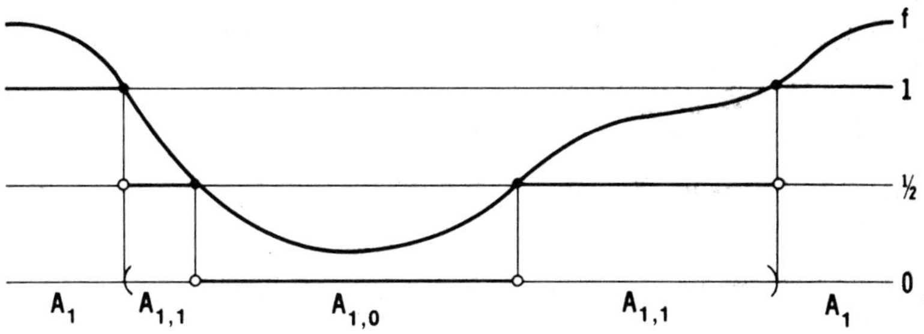

FIG. 7.1.

Exercises

(7-1) Show that any σ-algebra $\mathfrak{M}$ in a set X is either finite or uncountable. Hint: Observe that finite families of sets generate finite σ-algebras, and show that if $\mathfrak{M}$ is infinite, it contains an infinite family of nonempty, disjoint sets.

(7-2) If $\{A_n\}$ is a sequence of sets, let

$$\liminf A_n = \bigcup_{n=1}^{\infty} \bigcap_{k=n}^{\infty} A_k$$

$$\limsup A_n = \bigcap_{n=1}^{\infty} \bigcup_{k=n}^{\infty} A_k$$

If

$$A = \liminf A_n = \limsup A_n$$

we shall say that the sequence $\{A_n\}$ converges to A and write $\lim A_n = A$ or $A_n \to A$. Show that

(i) $\liminf A_n \subset \limsup A_n$

(ii) $(A_n \subset A_{n+1},\ n = 1, 2, \ldots) \Rightarrow \lim A_n = \bigcup_{n=1}^{\infty} A_n$

(iii) $(A_{n+1} \subset A_n,\ n = 1, 2, \ldots) \Rightarrow \lim A_n = \bigcap_{n=1}^{\infty} A_n$

(7-3) Let $\{A_n\}$ be a sequence of subsets of a set X. Show that

(i) $A = \liminf A_n \iff \chi_A = \liminf \chi_{A_n}$

(ii) $A = \limsup A_n \iff \chi_A = \limsup \chi_{A_n}$

(iii) $A = \lim A_n \iff \chi_A = \lim \chi_{A_n}$

(7-4)$^+$ Let X be a set. A family $\mathfrak{A} \subset \exp X$ is called an algebra in X whenever

(a) $X \in \mathfrak{A}$

(b) $A \in \mathfrak{A} \Rightarrow X - A \in \mathfrak{A}$

(c) $A, B \in \mathfrak{A} \Rightarrow A \cup B \in \mathfrak{A}$.

(i) Show that any algebra in X is closed with respect to the formation of finite unions, finite intersections, and set differences.

(ii) Prove that the intersection of any nonempty family of algebras in X is again an algebra in X.

(iii) Let $\mathfrak{C} \subset \exp X$. Show that there is the smallest algebra in X containing $\mathfrak{C}$. This algebra is called the algebra in X generated by $\mathfrak{C}$.

(7-5)$^{*+}$ Let X be a set, and let $\mathfrak{C} \subset \exp X$ satisfy the following conditions:

(a) $X \in \mathfrak{C}$

(b) $A, B \in \mathfrak{C} \Rightarrow A \cup B,\ A \cap B \in \mathfrak{C}$.

Let $\mathfrak{A}$ be the collection of all disjoint unions

$$\bigcup_{i=1}^{n} (A_i - B_i)$$

where $A_i, B_i \in \mathfrak{C}$ and $B_i \subset A_i$, $i = 1, \ldots, n$.

(i) Show that $\mathfrak{A}$ is the algebra in X generated by $\mathfrak{C}$. Hint: Read the proof of Lemma (18.8).

(ii) Observe that the families of closed and open subsets of a topological space satisfy conditions (a) and (b).

(7-6)$^+$ Let X be a set, $\mathfrak{C} \subset \exp X$, and let $\mathfrak{A}$ be the algebra generated by $\mathfrak{C}$ [see (7-4)(iii)].

(i) Let $\mathfrak{M}_0 = \mathfrak{A}$. Given an ordinal $\alpha < \Omega$, assume that the families $\mathfrak{M}_\beta$ have been already defined for all ordinals $\beta < \alpha$. Let $\mathfrak{M}_\alpha$ consist of all sets $A \subset X$ for which there is a sequence $\{A_n\} \subset \bigcup_{\beta<\alpha} \mathfrak{M}_\beta$ such that $A_n \to A$ [see (7-2)]. Show that $\mathfrak{M} = \bigcup_{\alpha<\Omega} \mathfrak{M}_\alpha$ is the σ-algebra in X generated by $\mathfrak{C}$. <u>Hint</u>: Observe that $\mathfrak{M}_\alpha$ is an algebra in X for each ordinal $\alpha < \Omega$ and that $\mathfrak{M}_\alpha \subset \mathfrak{M}_\beta$ whenever $\alpha < \beta < \Omega$.

(ii) Let $\mathfrak{D} \subset \exp X$ satisfy the following conditions:

(a) $\mathfrak{A} \subset \mathfrak{D}$

(b) If $\{A_n\} \subset \mathfrak{D}$ is a <u>monotone</u> sequence, then $\lim A_n$ belongs to $\mathfrak{D}$.

Show that $\mathfrak{M} \subset \mathfrak{D}$. <u>Hint</u>: Use the fact that $\mathfrak{M}_\alpha$ is an algebra in X for each ordinal $\alpha < \Omega$, and that

$$\lim A_n = \bigcap_{n=1}^{\infty} \bigcup_{m=n}^{\infty} \left(\bigcup_{k=n}^{m} A_k \right)$$

for each sequence $\{A_n\} \subset \exp X$.

<u>(7-7)</u>$^+$ Let X be a set and $\mathfrak{C} \subset \exp X$. Denote by $\mathfrak{A}$ and $\mathfrak{M}$, respectively, the algebra [see (7-4)] and the σ-algebra generated by $\mathfrak{C}$. Prove

(i) If $A \in \mathfrak{A}$, then there is a <u>finite</u> family $\mathfrak{C}_0 \subset \mathfrak{C}$ such that A belongs to the algebra generated by $\mathfrak{C}_0$. <u>Hint</u>: Show that the union of all algebras generated by finite subfamilies of $\mathfrak{C}$ is an algebra in X containing $\mathfrak{C}$.

(ii) If $A \in \mathfrak{M}$, then there is a <u>countable</u> family $\mathfrak{C}_1 \subset \mathfrak{C}$ such that A belongs to the σ-algebra generated by $\mathfrak{C}_1$. <u>Hint</u>: Show that the union of all σ-algebras generated by countable subfamilies of $\mathfrak{C}$ is a σ-algebra in X containing $\mathfrak{C}$.

<u>(7-8)</u> Let X be a set, and let $\mathfrak{M}$ consist of all $A \subset X$ for which either A or X - A is countable.

(i) Show that $\mathfrak{M}$ is a σ-algebra in X generated by the family $\{(x) : x \in X\}$.

(ii) Suppose that X is uncountable and consider the measurable space $(X, \mathfrak{M})$. Show that there is a nonmeasurable function f on X for which $|f|$ is measurable.

(7-9) Let $(X,\mathfrak{M})$ be a measurable space, and let D be a dense subset of $\mathbf{R}$. Let f be a function on X such that the set $\{x \in X : f(x) > a\}$ is measurable for every $a \in D$. Show that f is a measurable function.

(7-10)+ Let X be a set and $\mathfrak{C} \subset \exp X$. Let $\mathfrak{A}$ and $\mathfrak{M}$ denote, respectively, the algebra and the σ-algebra generated by $\mathfrak{C}$. Prove

(i) If $\|\mathfrak{C}\| \geq \aleph_0$, then $\|\mathfrak{A}\| = \|\mathfrak{C}\|$. Hint: Use (7-7)(i).

(ii) If $\|\mathfrak{C}\| \geq \aleph_0$, then $\|\mathfrak{M}\| \geq c$. Hint: Use the hint from (7-1).

(iii) If $\|\mathfrak{C}\| \leq c$, then $\|\mathfrak{M}\| \leq c$. Hint: Use (7-7)(ii) and (1-9)(iv).

Conclude from (iii) that if $\mathfrak{B}$ is the Borel σ-algebra in a second countable topological space (see Ref. 16, Chapter 1, p. 48), then $\|\mathfrak{B}\| \leq c$.

(7-11)+ A half-open cube K in $\mathbf{R}^m$, $m \geq 1$ an integer, is the m-fold cartesian product of an interval $[a,b)$ with $a,b \in \mathbf{R}$ and $a < b$. The cube K is called rational if a and b are rational numbers.

(i) Show that the Borel σ-algebra $\mathfrak{B}$ in $\mathbf{R}^m$ is generated by the family of all rational half-open cubes in $\mathbf{R}^m$. Hint: Use dyadic rationals to show that every open subset of $\mathbf{R}^m$ is a union of rational half-open cubes in $\mathbf{R}^m$.

(ii) Deduce from (i) that $\|\mathfrak{B}\| = c$. Hint: Use (7-10)(iii).

(7-12)*+ Let X be a topological space. The smallest σ-algebra in X with respect to which all functions from $C(X)$ are measurable is called the Baire σ-algebra in X, denoted by $\mathfrak{B}_o$. The elements of $\mathfrak{B}_o$ are called the Baire sets. Show that

(i) $\mathfrak{B}_o$ is generated by the family $\{f^{-1}(0) : f \in C(X)\}$. Hint: If $a \in \mathbf{R}$, then

$$\{x \in X : f(x) \geq a\} = g^{-1}(0)$$

where $g = (f - a)^-$.

(ii) $\mathfrak{B}_o \subset \mathfrak{B}$ [see (7.4)].

(iii) A function f on X is $\mathfrak{B}_o$-measurable iff it is a Baire function [see (3-13)]. Hint: Observe that if f is a function on X, $A = f^{-1}(0)$, and $g_n = 1 - (n|f|) \wedge 1$, then $g_n \searrow \chi_A$. Use this, (7-5), (3-14)(ii),

(7-6)(i), (7-3)(iii), (3-13)(ii), and (7.29) to show that all $\mathfrak{B}_o$-measurable functions are Baire functions. The converse is easy.

(7-13)+ Let X be a topological space.

(i) If $f \in C(X)$, show that $f^{-1}(0)$ is a closed G_δ set (see Chapter 1, Section B). Hint: Observe that

$$f^{-1}(0) = \bigcap_{n=1}^{\infty} \{x \in X : |f(x)| < \tfrac{1}{n}\}$$

(ii) Let X be normal, and let $F \subset X$ be a closed G_δ set. Show that there is $f \in C(X)$ such that $F = f^{-1}(0)$. Hint: Let $F = \bigcap_{n=1}^{\infty} G_n$, where $G_1, G_2, \ldots$ are open subsets of X. Using the Urysohn lemma, choose $f_n \in C(X)$, $n = 1, 2, \ldots$, such that $0 \leq f_n \leq 1$, $f_n(x) = 0$ if $x \in A$, and $f_n(x) = 1$ if $x \in X - G_n$. Show that the function $f = \Sigma_{n=1}^{\infty} 2^{-n} f_n$ has the desired properties.

(iii) Conclude from (i), (ii), and (7-12)(i) that in a normal space X the Baire σ-algebra $\mathfrak{B}_o$ is generated by the family of all closed G_δ subsets of X.

(iv) Conclude from (iii) that in a perfectly normal space X [see (1-12)] the Borel and Baire σ-algebras coincide.

Note: In spite of (iii), in a normal space which is not perfectly normal there may be closed Baire sets which are not G_δ (see Ref. 20, example 3.2).

(7-14) Let $X = W$ or $X = W^-$ [see Chapter 1, Section B], and let $\mathfrak{B}$ and $\mathfrak{B}_o$ denote, respectively, the Borel and the Baire σ-algebras in X. Show that

(i) $A \in \mathfrak{B}_o$ iff A or X - A is a countable subset of W. Hint: Use (1-15)(ii).

(ii) $\mathfrak{B}_o \subsetneqq \mathfrak{B}$.

(7-15)* Let X be a nonempty set, and let $\mathfrak{A}$ be a finite algebra in X. Prove the following:

(i) $\mathfrak{A}$ is a σ-algebra in X.

(ii) There is an integer $n \geq 1$ such that $\|\mathfrak{A}\| = 2^n$. Hint: Observe that $\mathfrak{A}$ is generated by the nonempty elements of $\mathfrak{A}$ which are minimal with respect to the inclusion.

(iii) If $Y = \{1, \ldots, n\}$ where n is the integer from (ii), then there is a surjective (i.e., onto) map $f : X \to Y$ which is $(\mathfrak{A}, \exp Y)$-measurable.

8. MEASURES AND THE ABSTRACT LEBESGUE INTEGRAL

When we defined a measurable space $(X,\mathfrak{M})$ [see (7.6)], we mentioned that the measurable sets are those subsets of X to which a "measure" can be ascribed. In this chapter we shall define precisely what we mean by a measure of a set and we shall derive some basic properties of measures. Intuitively speaking, one can picture a measure as a far-reaching generalization of such familiar concepts as length, surface area, volume, mass, and electric charge.

(8.1) Definition. Let $(X,\mathfrak{M})$ be a measurable space. A nonnegative function μ on $\mathfrak{M}$ is called a measure on $\mathfrak{M}$ if the following conditions hold:

(i) $\mu(\emptyset) = 0$

(ii) If $\{A_n\} \subset \mathfrak{M}$ is a countable disjoint family, then

$$\mu(\bigcup_n A_n) = \sum_n \mu(A_n)$$

Notice that condition (i) serves only one purpose. Namely, it eliminates from our considerations the trivial case when $\mu(A) = +\infty$ for all $A \in \mathfrak{M}$. Indeed, by (ii),

$$\mu(\emptyset) = \mu(\emptyset \cup \emptyset) = \mu(\emptyset) + \mu(\emptyset) = 2\mu(\emptyset)$$

and so if $\mu(\emptyset) > 0$, we have $\mu(\emptyset) = +\infty$; this implies

$$\mu(A) = \mu(A \cup \emptyset) = \mu(A) + \mu(\emptyset) = +\infty$$

for each $A \in \mathfrak{M}$.

It is unfortunate that good intuitive examples of measures are not readily available. The most important measures have to be laboriously constructed and the constructions are usually far from simple. In exercises (8-1) and (8-2) we shall give some examples of measures. However, they all will be too simple to illustrate well the full complexity which is involved in the notion of a measure. Nontrivial examples of measures will be given in Chapter 10.

(8.2) Definition. A measure space is a triple $(X,\mathfrak{M},\mu)$, where X is a set, $\mathfrak{M}$ is a σ-algebra in X, and μ is a measure on $\mathfrak{M}$.

Throughout this chapter we shall assume that $(X,\mathfrak{M},\mu)$ is a given measure space.

(8.3) Proposition. If $A, B \in \mathfrak{M}$ and $A \subset B$, then $\mu(A) \leq \mu(B)$. If, in addition, $\mu(A) < +\infty$, then

$$\mu(B - A) = \mu(B) - \mu(A)$$

Proof. Since $B = A \cup (B - A)$, we have

$$\mu(B) = \mu(A) + \mu(B - A)$$

and the proposition follows. ///

(8.4) Proposition. If $\{A_n\} \subset \mathfrak{M}$ is any countable family (not necessarily disjoint), then

$$\mu(\bigcup_n A_n) \leq \sum_n \mu(A_n)$$

Proof. Let $B_n = A_n - \bigcup_{i=1}^{n-1} A_i$, $n = 1, 2, \ldots$. Then $\{B_n\} \subset \mathfrak{M}$ is a disjoint countable family and $\bigcup_n B_n = \bigcup_n A_n$. Thus by (8.3),

$$\mu(\bigcup_n A_n) = \sum_n \mu(B_n) \leq \sum_n \mu(A_n) \quad ///$$

(8.5) Proposition. If $\{A_n\} \subset \mathfrak{M}$ and $A_n \nearrow A$, then $\mu(A_n) \nearrow \mu(A)$.

Proof. By (8.3),

$$\mu(A_n) \leq \mu(A_{n+1}) \leq \mu(A)$$

$n = 1, 2, \ldots$, and so the proposition holds if $\mu(A_n) = +\infty$ for some n. Hence, assume that $\mu(A_n) < +\infty$ for all n. Since

$$A = A_1 \cup \bigcup_{n=1}^{\infty} (A_{n+1} - A_n)$$

is a disjoint union, using (8.3), we obtain

$$\mu(A) = \mu(A_1) + \sum_{n=1}^{\infty} [\mu(A_{n+1}) - \mu(A_n)]$$
$$= \lim_{N\to+\infty} \{\mu(A_1) + \sum_{n=1}^{N-1} [\mu(A_{n+1}) - \mu(A_n)]\}$$
$$= \lim_{N\to+\infty} \mu(A_N) \quad ///$$

(8.6) Proposition. If $\{A_n\} \subset \mathfrak{M}$, $A_n \searrow A$, and $\mu(A_1) < +\infty$, then $\mu(A_n) \searrow \mu(A)$.

Proof. Using (8.3) and (8.5) we obtain

$$\mu(A_1) - \mu(A) = \mu(A_1 - \bigcap_{n=1}^{\infty} A_n) = \mu[\bigcup_{n=1}^{\infty} (A_1 - A_n)]$$
$$= \lim \mu(A_1 - A_n) = \mu(A_1) - \lim \mu(A_n)$$

The proposition follows. ///

We note that Proposition (8.6) is generally false if the assumption $\mu(A_1) < +\infty$ is omitted [see (8-3)].

(8.7) Definition. The family $\mathfrak{M}_o = \{A \in \mathfrak{M} : \mu(A) = 0\}$ is called the null ideal. The elements of $\mathfrak{M}_o$ are called the null sets.

If we need to be more specific, we shall say the μ-null ideal in $\mathfrak{M}$ instead of the null ideal. Similarly, instead of a null set we may say a μ-null set in $\mathfrak{M}$.

From (8.3) and (8.4) it follows that

(8.8) $(A \in \mathfrak{M}_o, B \in \mathfrak{M}) \Rightarrow A \cap B \in \mathfrak{M}_o$.

(8.9) If $\{A_n\} \subset \mathfrak{M}_o$ is a countable family, then $\bigcup_n A_n \in \mathfrak{M}_o$.

(8.10) Definition. A measure μ on a σ-algebra $\mathfrak{M}$ is called complete whenever all subsets of every null set are measurable.

Using (8.8) we obtain

(8.11) If μ is complete, then

$$(A \in \mathfrak{M}_o, B \subset X) \Rightarrow A \cap B \in \mathfrak{M}_o$$

(8.12) Definition. A measure μ on a σ-algebra $\mathfrak{M}$ is called saturated whenever the following condition holds: If $A \subset X$ and $A \cap B \in \mathfrak{M}$ for all $B \in \mathfrak{M}$ for which $\mu(B) < +\infty$, then $A \in \mathfrak{M}$.

Being consistent with our terminology from Chapter 1, Section D, a measure is called finite if its values are real numbers. Clearly, a measure μ on a σ-algebra $\mathfrak{M}$ in X is finite iff $\mu(X) < +\infty$. Since $A = A \cap X$ for each $A \subset X$, every finite measure is saturated. We shall prove more, but first we need additional definitions.

(8.13) Definition. A set $A \in \mathfrak{M}$ is called σ-finite if there are sets $A_n \in \mathfrak{M}$ such that $\mu(A_n) < +\infty$, $n = 1, 2, \ldots$, and $A = \bigcup_{n=1}^{\infty} A_n$.

Again, to be more specific, we shall sometimes say μ-σ-finite set instead of σ-finite set.

(8.14) Definition. A measure μ on a σ-algebra $\mathfrak{M}$ is called σ-finite if each set from $\mathfrak{M}$ is σ-finite.

Obviously, every finite measure is σ-finite. From (8.3) we obtain

(8.15) A measure μ is σ-finite iff the set X is σ-finite.

(8.16) Proposition. Every σ-finite measure is saturated.

Proof. Suppose that μ is a σ-finite measure. Then there are sets $X_n \in \mathfrak{M}$ such that $\mu(X_n) < +\infty$, $n = 1, 2, \ldots$, and $X = \bigcup_{n=1}^{\infty} X_n$. Let $A \subset X$ be such that $A \cap B \in \mathfrak{M}$ for all $B \in \mathfrak{M}$ for which $\mu(B) < +\infty$. Then $A \in \mathfrak{M}$, for

$$A = A \cap X = \bigcup_{n=1}^{\infty} (A \cap X_n) \quad ///$$

We shall close this chapter by defining the abstract Lebesgue integral with respect to a measure μ.

If f is a nonnegative simple function on X [see (7.25)] and if

$$f = \sum_{i=1}^{n} a_i \chi_{A_i}$$

is its canonical form [see (7.27)], let

$$\int f \, d\mu = \sum_{i=1}^{n} a_i \mu(A_i)$$

Since $f \geq 0$, we have $a_i \geq 0$ for $i = 1, \ldots, n$, and the sum on the right side has meaning. It follows from (7.28) that $\int f \, d\mu$ is a well-defined nonnegative extended real number.

If f is a nonnegative measurable function on X, let

$$\int f \, d\mu = \sup\{\int g \, d\mu : g \in \mathcal{S}, \ 0 \leq g \leq f\}$$

where $\mathcal{S}$ denotes the family of all simple functions on X.

For an arbitrary measurable function on X we shall give the following definition.

(8.17) Definition. A measurable function f on X is called Lebesgue integrable whenever $\int f^+ \, d\mu - \int f^- \, d\mu$ has meaning. The extended real number

$$\int f \, d\mu = \int f^+ \, d\mu - \int f^- \, d\mu$$

is then called the abstract Lebesgue integral of f with respect to μ.

Directly from this definition one can show that the abstract Lebesgue integral has properties remarkably similar to those of the Daniell and Bourbaki integrals. This is a standard procedure which can be found in almost every real-variable textbook. We shall, however, adopt a different method. Namely, we shall show in Chapter 12 that the abstract Lebesgue integral is a restriction of a certain Daniell integral [see (12.19)]. Thus using our previous results we shall immediately establish all basic properties of the abstract Lebesgue integral.

Exercises

(8-1)[+] Let $\mathfrak{M}$ be a σ-algebra in a set X. Show that the following functions μ on $\mathfrak{M}$ are measures.

(i) $\mu(A) = 0$ for all $A \in \mathfrak{M}$.

(ii) $\mu(\emptyset) = 0$ and $\mu(A) = +\infty$ for all nonempty $A \in \mathfrak{M}$.

(iii) $\mu(A) = 0$ or $\mu(A) = +\infty$ according to whether $A \in \mathfrak{M}$ is, respectively, countable or uncountable.

(iv) $\mu(A) = \|A\|$ or $\mu(A) = +\infty$ according to whether $A \in \mathfrak{M}$ is, respectively, finite or infinite. This measure is called the counting measure.

(v) Let $x_o \in X$, and for each $A \in \mathfrak{M}$, let

$$\mu(A) = \chi_A(x_o)$$

where χ_A is the characteristic function of A [see (7.19)]. This measure is called the Dirac measure at x_o .

(vi) Let f be a nonnegative function on X, and for each $A \in \mathfrak{M}$, let

$$\mu(A) = \Sigma\{f(x) : x \in A\}$$

This measure is called the weighted counting measure determined by f. Hint: Use (5-9)(v).

(8-2) Let X be a set and let $\aleph$ be an infinite cardinal. Denote by $\mathfrak{M}$ the family of all $A \subset X$ such that either $\|A\| \leq \aleph$ or $\|X - A\| \leq \aleph$. For $A \in \mathfrak{M}$ set $\mu(A) = \nu(A) = 0$ if $\|A\| \leq \aleph$, and $\mu(A) = 1$, $\nu(A) = +\infty$ if $\|A\| > \aleph$. Show that

(i) μ and ν are complete measures on $\mathfrak{M}$.

(ii) ν is saturated iff $\|X\| \leq \aleph$.

(8-3) Let $X = \{1,2,\ldots\}$, and let μ be the counting measure on exp X. Set $A_n = \{k \in X : k \geq n\}$, and observe that $\mu(A_n) = +\infty$, $n = 1, 2, \ldots$, and $\mu(\bigcap_{n=1}^{\infty} A_n) = 0$.

(8-4) Let X be an infinite set, and let $\mathfrak{M} = \{\emptyset, X\}$. Denote by μ the counting measure on exp X and by ν the measure on exp X which is identically equal to zero. Show that

(i) μ restricted to $\mathfrak{M}$ is a complete but not saturated measure.

(ii) ν restricted to $\mathfrak{M}$ is a saturated but not complete measure.

(iii) μ on exp X is σ-finite iff X is countable.

Give an example of a measure which is neither complete nor saturated.

(8-5)[+] Let $\mathfrak{M}$ be a σ-algebra in a set X. A nonnegative function μ on $\mathfrak{M}$ is called a finitely additive measure on $\mathfrak{M}$ if

(a) $\mu(\emptyset) = 0$

(b) $(A, B \in \mathfrak{M},\ A \cap B = \emptyset) \Rightarrow \mu(A \cup B) = \mu(A) + \mu(B)$.

To make the distinction between finitely additive measures and measures sharper, the measures from Definition (8.1) are sometimes called the σ-additive measures.

Show that a finitely additive measure μ on $\mathfrak{M}$ is a σ-additive measure on $\mathfrak{M}$ whenever either of the following conditions is satisfied.

(i) For every disjoint sequence $\{A_n\} \subset \mathfrak{M}$,

$$\mu\left(\bigcup_{n=1}^{\infty} A_n\right) \leq \sum_{n=1}^{\infty} \mu(A_n)$$

(ii) For every sequence $\{A_n\} \subset \mathfrak{M}$ with $A_n \nearrow A$,

$$\lim \mu(A_n) = \mu(A)$$

(iii) For every sequence $\{A_n\} \subset \mathfrak{M}$ with $A_n \searrow \emptyset$,

$$\lim \mu(A_n) = 0$$

(iv) There is a finite σ-additive measure ν on $\mathfrak{M}$ such that $\mu(A) \leq \nu(A)$ for each $A \in \mathfrak{M}$. Hint: Use (iii) and (8.6).

(8-6)* Let $(X, \mathfrak{M}, \mu)$ be a measure space. Show that

(i) If μ is finite, then the set $\{\mu(A) : A \in \mathfrak{M}\}$ is either finite or uncountable. Hint: Apply the hint in (7-1) to exhibit a disjoint family $\{A_n\}_{n=1}^{\infty} \subset \mathfrak{M}$ such that $\mu(A_n) > \sum_{k=n+1}^{\infty} \mu(A_k)$ for each $n = 1, 2, \ldots$.

(ii) The finiteness of μ is essential in (i). Hint: Consider the measure μ from (8-3).

(8-7)$^+$ Let $(X,\mathfrak{M},\mu)$ be a measure space, and let $\{A_n\} \subset \mathfrak{M}$. Prove

(i) $\mu(\liminf A_n) \leq \liminf \mu(A_n)$.

(ii) If $\mu(\bigcup_{n=1}^{\infty} A_n) < +\infty$, then

$$\mu(\limsup A_n) \geq \limsup \mu(A_n)$$

(iii) If $\mu(\bigcup_{n=1}^{\infty} A_n) < +\infty$ and $A_n \to A$, then $\mu(A) = \lim \mu(A_n)$.

For the definition of the symbols see (7-2).

(8-8)$^+$ Egoroff's theorem. Let $(X,\mathfrak{M},\mu)$ be a measure space, $A \in \mathfrak{M}$, and let f and f_n, $n = 1, 2, \ldots$, be finite measurable functions on X such that $f_n(x) \to f(x)$ for each $x \in A$. If $\mu(A) < +\infty$, then to every $\varepsilon > 0$ there is an $M \in \mathfrak{M}$ such that $\mu(M) < \varepsilon$ and $f_n \to f$ uniformly on $A - M$.

(i) Prove Egoroff's theorem. Hint: For $p,q = 1, 2, \ldots$, let $A_q^p = \{x \in A : |f_n(x) - f(x)| \geq 1/p,\ n \geq q\}$, and show that $\lim_{q\to+\infty} \mu(A_q^p) = 0$, $p = 1, 2, \ldots$. Then choose q_p such that $\mu(A_{q_p}^p) < \varepsilon 2^{-p}$ and set $M = \bigcup_{p=1}^{\infty} A_{q_p}^p$.

(ii) Show that the assumption $\mu(A) < +\infty$ is essential in the Egoroff theorem. Hint: Consider the measure space from (8-3), and let $A = X$ and $f_n(x) = x/n$, $x \in X$, $n = 1, 2, \ldots$.

(8-9)$^+$ Let $(X,\mathfrak{M},\mu)$ be a measure space. Denote by $\bar{\mathfrak{M}}$ the system of all sets $A \subset X$ for which there is a set $B \in \mathfrak{M}$ such that $(A - B) \cup (B - A)$ is a subset of a μ-null set [see (8.7)]. If $A \in \bar{\mathfrak{M}}$, find $B \in \mathfrak{M}$ such that $(A - B) \cup (B - A)$ is a subset of a μ-null set, and let $\bar{\mu}(A) = \mu(B)$. Show that

(i) $\bar{\mathfrak{M}}$ is a σ-algebra in X containing $\mathfrak{M}$. Hint: Use (8.8) and (8.9).

(ii) $\bar{\mu}$ is a well-defined complete measure on $\bar{\mathfrak{M}}$.

(iii) $\bar{\mu}(A) = \mu(A)$ for each $A \in \mathfrak{M}$.

(iv) $\bar{\mu}$ is σ-finite iff μ is σ-finite.

(v) Let $\mathfrak{M}'$ be a σ-algebra in X containing $\mathfrak{M}$ and let μ' be a complete measure on $\mathfrak{M}'$ such that $\mu'(A) = \mu(A)$ for each $A \in \mathfrak{M}$. Then $\bar{\mathfrak{M}} \subset \mathfrak{M}'$ and $\mu'(A) = \bar{\mu}(A)$ for each $A \in \bar{\mathfrak{M}}$.

Thus $\bar{\mu}$ is the unique minimal complete extension of μ. It is called the <u>completion</u> of the measure μ.

<u>(8-10)</u>$^+$ Let $(X,\mathfrak{M},\mu)$ be a measure space. Denote by $\mathfrak{M}^\wedge$ the system of all $A \subset X$ such that $A \cap B \in \mathfrak{M}$ whenever $B \in \mathfrak{M}$ and $\mu(B) < +\infty$. For $A \in \mathfrak{M}^\wedge$ let $\mu^\wedge(A) = \mu(A)$ if $A \in \mathfrak{M}$, and $\mu(A) = +\infty$ otherwise. Show that

(i) $\mathfrak{M}^\wedge$ is a σ-algebra in X containing $\mathfrak{M}$.

(ii) $\mu^\wedge$ is a saturated measure on $\mathfrak{M}^\wedge$.

(iii) If μ is complete, so is $\mu^\wedge$.

(iv) If μ is saturated, then $\mathfrak{M}^\wedge = \mathfrak{M}$.

The measure $\mu^\wedge$ on $\mathfrak{M}^\wedge$ is called the <u>saturation</u> of the measure μ.

<u>Note</u>: In general, the saturation μ does not have the pleasant uniqueness property analogous to that of the completion $\bar{\mu}$ [see (8-9)(v)]. If $\mathfrak{M}'$ is a σ-algebra in X containing $\mathfrak{M}$ and μ' is a saturated measure on $\mathfrak{M}'$ such that $\mu'(A) = \mu(A)$ for each $A \in \mathfrak{M}$, then it may happen that $\mathfrak{M}^\wedge \not\subset \mathfrak{M}'$; even on $\mathfrak{M}^\wedge \cap \mathfrak{M}'$ the measures $\mu^\wedge$ and μ' may not coincide [see (10-10)]. However, there is an important special case in which the saturation $\mu^\wedge$ is the unique minimal saturated extension of μ [see (9-6)].

Let $(X,\mathfrak{M})$ be a measurable space. A measure μ on $\mathfrak{M}$ is called <u>diffused</u> if $\mu(\{x\}) = 0$ for each $x \in X$ for which $\{x\} \in \mathfrak{M}$.

<u>(8-11)</u>* Let $(X,\mathfrak{M},\mu)$ be a measure space and let $\{x\} \in \mathfrak{M}$ for each $x \in X$. Show that

(i) There are two measures ι and ν on $\mathfrak{M}$ such that ι is diffused, ν is a weighted counting measure, and

$$\mu(A) = \iota(A) + \nu(A)$$

for each $A \in \mathfrak{M}$. <u>Hint</u>: Let $f(x) = \mu(\{x\})$ for each $x \in X$, and let ν be the weighted counting measure on $\mathfrak{M}$ determined by f. If $A \in \mathfrak{M}$ is μ-σ-finite, find disjoint sets $A_n \in \mathfrak{M}$ such that $\mu(A_n) < +\infty$, $n = 1, 2, \ldots$, and $A = \bigcup_{n=1}^\infty A_n$; then set

$$\iota(A) = \sum_{n=1}^{\infty} [\mu(A_n) - \nu(A_n)]$$

If $A \in \mathfrak{M}$ is not μ-σ-finite, set $\iota(A) = +\infty$.

(ii) The measure ν is determined uniquely by μ, while the measure ι is generally not unique. <u>Hint</u>: In an uncountable space X add the measure from (8-1)(iii) to the counting measure.

(iii) If μ is σ-finite, then both measures ι and ν are determined uniquely by μ.

(iv) If X is countable, then $\mathfrak{M} = \exp X$ and μ is a weighted counting measure.

(v) Let μ be the weighted counting measure determined by a function f on X. Then μ is σ-finite iff f is finite and $\{x \in X : f(x) > 0\}$ is countable. <u>Hint</u>: Use (5-8).

(vi) Let μ be a weighted counting measure. Then μ is complete and saturated iff $\mathfrak{M} = \exp X$. <u>Hint</u>: Let μ be determined by a function f on X. Observe that if μ is saturated, then $\exp\{x \in X : f(x) > 0\} \subset \mathfrak{M}$.

<u>(8-12)</u> Let $(X, \mathfrak{M})$ be a measurable space. Prove the following:

(i) Let μ_n, n = 1, 2, ..., be measures on $\mathfrak{M}$, and let

$$\mu(A) = \sum_{n=1}^{\infty} \mu_n(A)$$

for each $A \in \mathfrak{M}$. Then μ is a measure on $\mathfrak{M}$.

(ii) Let μ be a σ-finite measure on $\mathfrak{M}$. Then there are finite measures μ_n on $\mathfrak{M}$ such that

$$\mu(A) = \sum_{n=1}^{\infty} \mu_n(A)$$

for each $A \in \mathfrak{M}$. <u>Hint</u>: Choose a disjoint sequence $\{X_n\} \subset \mathfrak{M}$ so that $\mu(X_n) < +\infty$, n = 1, 2, ..., and $X = \bigcup_{n=1}^{\infty} X_n$. For $A \in \mathfrak{M}$ and n = 1, 2, ..., let $\mu_n(A) = \mu(A \cap X_n)$.

<u>(8-13)</u>* Let $(X, \mathfrak{M}, \mu)$ be a measure space, and let f be a measurable function on X. Show that

(i) If μ is the Dirac measure at x_o, then f is Lebesgue integrable and

$$\int f \, d\mu = f(x_o)$$

(ii) If μ is the counting measure, then f is Lebesgue integrable iff Σf is defined [see (5-9)] and

$$\int f \, d\mu = \Sigma f$$

<u>(8-14)</u>† Suppose we have a bag of change consisting of, say, pennies, nickles, dimes, and quarters. To find out how much money is in the bag we have to count the change and there are two obvious ways how we can do the counting.

(i) We take out of the bag coin after coin and simply add their values as we go.

(ii) We take all coins out of the bag at once and make piles from the coins of the same denominations. We count the number of coins in each pile, multiply it by the denomination corresponding to the pile, and then we add all the resulting numbers.

Each way of counting corresponds to either Riemann or Lebesgue integration. Decide which corresponds to which and carefully justify your assertion. <u>Hint</u>: Let X be the set of all coins, and let μ be the counting measure on exp X.

A family $\mathfrak{C}$ of sets is called <u>star-countable</u> if for each $A \in \mathfrak{C}$, the family $\{B \in \mathfrak{C} : A \cap B \neq \emptyset\}$ is countable.

<u>(8-15)</u> Let $\mathfrak{C}$ be a star-countable family of sets. For A and B in $\mathfrak{C}$ write $A \sim B$ iff there are sets $C_0, \ldots, C_n$ in $\mathfrak{C}$ such that $C_0 = A$, $C_n = B$, and $C_{i-1} \cap C_i \neq \emptyset$ for $i = 1, \ldots, n$. Show that $\sim$ is an equivalence relation on $\mathfrak{C}$, and denote by $\mathfrak{C}_\alpha$, $\alpha \in T$, the corresponding equivalence classes. Prove the following:

† J. B. Diaz contributed this problem; it was presented to the author during his Ph.D. qualifying examination.

(i) Each $\mathfrak{C}_\alpha$ is countable. Hint: Choose $A \in \mathfrak{C}$ and let

$$\mathfrak{A}_1 = \{B \in \mathfrak{C} : A \cap B \neq \emptyset\}$$
$$A_1 = \bigcup\{B : B \in \mathfrak{A}_1\}$$

If A_n has been defined, let

$$\mathfrak{A}_{n+1} = \{B \in \mathfrak{C} : A_n \cap B \neq \emptyset\}$$
$$A_{n+1} = \bigcup\{B : B \in \mathfrak{A}_{n+1}\}$$

Show that each $\mathfrak{A}_n$ is countable, and observe that $\bigcup_{n=1}^{\infty} \mathfrak{A}_n$ is the equivalence class determined by A.

(ii) For $\alpha \in T$, let $A_\alpha = \bigcup\{A : A \in \mathfrak{C}_\alpha\}$. Then $\{A_\alpha : \alpha \in T\}$ is a disjoint family.

(8-16)+ Let $(X, \mathfrak{M}, \mu)$ be a measure space with a σ-finite measure μ, and let $\mathfrak{C} \subset \mathfrak{M}$. Show that

(i) If $\mathfrak{C}$ is a disjoint family, then $\{A \in \mathfrak{C} : \mu(A) > 0\}$ is a countable family. Hint: If μ is finite, use (5-8). Then apply (8-12)(ii).

(ii) Statement (i) remains correct if $\mathfrak{C}$ is a star-countable family. Hint: Use (i) and (8-15).

9. MEASURES IN A TOPOLOGICAL SPACE

In topological spaces we shall study measures whose values on an arbitrary measurable set can be approximated by values on compact subsets and open supersets.

Throughout this chapter we shall assume that X is a locally compact Hausdorff space. By $\mathfrak{F}$ and $\mathfrak{G}$ we shall denote the families of all compact and all open subsets of X, respectively.

A set $F \subset X$ is called σ-compact if it is a countable union of compact subsets of X. The family of all σ-compact subsets of X is denoted by $\mathfrak{F}_\sigma$. We shall denote by $\mathfrak{G}_\delta$ the family of all G_δ subsets of X (see Chapter 1, Section B).

(9.1) Definition. Let $\mathfrak{M}$ be a σ-algebra in X. A measure μ on $\mathfrak{M}$ is called regular whenever the following conditions hold:

(i) $\mathfrak{G} \subset \mathfrak{M}$

(ii) $F \in \mathfrak{F} \Rightarrow \mu(F) < +\infty$

(iii) $G \in \mathfrak{G} \Rightarrow \mu(G) = \sup\{\mu(F) : F \in \mathfrak{F},\ F \subset G\}$

(iv) $A \in \mathfrak{M} \Rightarrow \mu(A) = \inf\{\mu(G) : G \in \mathfrak{G},\ A \subset G\}$.

It follows from (i) that $\mathfrak{M}$ contains the Borel σ-algebra $\mathfrak{B}$ in X [see (7.4)]. In particular, $\mathfrak{F} \subset \mathfrak{M}$ and (ii) and (iii) have meaning.

The regular measures are very important. For instance, we shall see in Chapter 13 that the regular measures in X which are complete and saturated are in one-to-one correspondence with the nonnegative linear functionals on $C_o(X)$.

Throughout this chapter we shall assume that $\mathfrak{M}$ is a σ-algebra in X and that μ is a regular measure on $\mathfrak{M}$.

(9.2) Lemma. If $A \in \mathfrak{M}$ and $\mu(A) < +\infty$, then

$$\mu(A) = \sup\{\mu(F) : F \in \mathfrak{F},\ F \subset A\}$$

<u>Proof</u>. Choose $\epsilon > 0$. By (9.1)(iv), there is a $G \in \mathfrak{G}$ such that

$$A \subset G \quad \text{and} \quad \mu(G) < \mu(A) + \frac{\epsilon}{2}$$

Thus $\mu(G - A) < \epsilon/2$ [see (8.3)], and using (9.1)(iv) again, we can find an $H \in \mathfrak{G}$ such that

$$G - A \subset H \quad \text{and} \quad \mu(H) < \frac{\epsilon}{2}$$

By (9.1)(iii) there is an $F \in \mathfrak{F}$ for which

$$F \subset G \quad \text{and} \quad \mu(F) > \mu(G) - \frac{\epsilon}{2}$$

Letting $E = F - H$, we have $E \in \mathfrak{F}$, $E \subset A$, and, by (8.3), also

$$\mu(E) = \mu(F) - \mu(H \cap F) > \mu(G) - \frac{\epsilon}{2} - \mu(H) > \mu(A) - \epsilon$$

The lemma follows from the arbitrariness of ϵ. ///

<u>(9.3)</u> <u>Corollary</u>. If $A \in \mathfrak{M}$ is σ-finite, then

$$\mu(A) = \sup\{\mu(F) : F \in \mathfrak{F},\ F \subset A\}$$

<u>Proof</u>. Let $A \in \mathfrak{M}$ be σ-finite. Then there are $A_n \in \mathfrak{M}$ such that $\mu(A_n) < +\infty$, $n = 1, 2, \ldots$, and $A = \bigcup_{n=1}^{\infty} A_n$. Replacing A_n by $\bigcup_{i=1}^{n} A_i$ if necessary, we may assume that $A_n \subset A_{n+1}$. By (8.5) and (9.2),

$$\begin{aligned} \mu(A) &= \sup_n \mu(A_n) = \sup_n [\sup\{\mu(F) : F \in \mathfrak{F},\ F \subset A_n\}] \\ &\le \sup\{\mu(F) : F \in \mathfrak{F},\ F \subset A\} \le \mu(A) \quad /// \end{aligned}$$

<u>(9.4)</u> <u>Lemma</u>. Let $A \in \mathfrak{M}$ be σ-finite, and let $\epsilon > 0$. Then there are sets $F \in \mathfrak{F}_\sigma$ and $G \in \mathfrak{G}$ such that

$$F \subset A \subset G \quad \text{and} \quad \mu(G - F) \le \epsilon$$

<u>Proof</u>. There are $A_n \in \mathfrak{M}$ such that $\mu(A_n) < +\infty$, $n = 1, 2, \ldots$, and $A = \bigcup_{n=1}^{\infty} A_n$. Using (9.1)(iv) and (9.2), we can find $F_n \in \mathfrak{F}$ and $G_n \in \mathfrak{G}$ so that $F_n \subset A_n \subset G_n$ and

$$\mu(F_n) > \mu(A_n) - \epsilon 2^{-n-1} \qquad \mu(G_n) < \mu(A_n) + \epsilon 2^{-n-1}$$

$n = 1, 2, \ldots$. Letting $F = \bigcup_{n=1}^{\infty} F_n$ and $G = \bigcup_{n=1}^{\infty} G_n$, we have $F \in \mathfrak{F}_\sigma$, $G \in \mathfrak{G}$, and $F \subset A \subset G$. It follows from (8.3) and (8.4) that

$$\mu(G - F) \leq \mu[\bigcup_{n=1}^{\infty} (G_n - F_n)] \leq \sum_{n=1}^{\infty} \mu(G_n - F_n) \leq \epsilon \sum_{n=1}^{\infty} 2^{-n} = \epsilon \;///$$

<u>(9.5)</u> <u>Corollary</u>. Let $A \in \mathfrak{M}$ be σ-finite. Then there are sets $F \in \mathfrak{F}_\sigma$ and $G \in \mathfrak{G}_\delta$ such that

$$F \subset A \subset G \quad \text{and} \quad \mu(G - F) = 0$$

<u>Proof</u>. By (9.4) there are sets $F_n \in \mathfrak{F}_\sigma$ and $G_n \in \mathfrak{G}$ such that

$$F_n \subset A \subset G_n \quad \text{and} \quad \mu(G_n - F_n) \leq \frac{1}{n}$$

$n = 1, 2, \ldots$. It suffices to let $F = \bigcup_{n=1}^{\infty} F_n$, $G = \bigcap_{n=1}^{\infty} G_n$, and apply (8.3). ///

We note that, in general, the condition of σ-finiteness cannot be omitted from Lemma (9.4) and Corollaries (9.3) and (9.5) [see (11-3)(i)].

<u>(9.6)</u> <u>Proposition</u>. Let $\mathfrak{D}$ be a directed system of open sets, and let $\mathfrak{D} \nearrow G_o$. Then

$$\mu(G_o) = \sup\{\mu(G) : G \in \mathfrak{D}\}$$

<u>Proof</u>. Choose $\alpha < \mu(G_o)$ and $F \in \mathfrak{F}$ such that $F \subset G_o$ and $\mu(F) > \alpha$. Since F is a compact subset of G_o and $\mathfrak{D} \nearrow G_o$, there is a $G \in \mathfrak{D}$ containing F. Thus $\mu(G) > \alpha$, and the proposition follows from the arbitrariness of α. ///

<u>(9.7)</u> <u>Proposition</u>. Let $\mathfrak{C}$ be a directed system of closed sets, and let $\mathfrak{C} \searrow F_o$. If $\mu(F) < +\infty$ for some $F \in \mathfrak{C}$, then

$$\mu(F_o) = \inf\{\mu(F) : F \in \mathfrak{C}\}$$

<u>Proof</u>. Let $C \in \mathfrak{C}$ be such that $\mu(C) < +\infty$. Choose $G_o \in \mathfrak{G}$ for which $C \subset G_o$ and $\mu(G_o) < +\infty$. Letting

$$\mathfrak{D} = \{G_o - F : F \in \mathfrak{C},\ F \subset C\}$$

we have $\mathfrak{D} \subset \mathfrak{G}$ and $\mathfrak{D} \nearrow G_o - F_o$. Thus by (8.3) and (9.6),

$$\begin{aligned}\mu(G_o) - \mu(F_o) = \mu(G_o - F_o) &= \sup\{\mu(G) : G \in \mathfrak{D}\}\\ &= \sup\{\mu(G_o) - \mu(F) : F \in \mathfrak{C},\ F \subset C\}\\ &= \mu(G_o) - \inf\{\mu(F) : F \in \mathfrak{C},\ F \subset C\}\end{aligned}$$

It follows that

$$\begin{aligned}\mu(F_o) &= \inf\{\mu(F) : F \in \mathfrak{C},\ F \subset C\}\\ &\geq \inf\{\mu(F) : F \in \mathfrak{C}\} \geq \mu(F_o) \quad ///\end{aligned}$$

The assumption that $\mu(F) < +\infty$ for some $F \in \mathfrak{C}$ cannot be omitted from Proposition (9.7). To see this, consider exercise (8-3) with X having the discrete topology [see (9-2)].

(9.8) Proposition. Let μ be complete. Then μ is saturated iff $A \in \mathfrak{M}$ whenever $A \cap F \in \mathfrak{M}$ for each $F \in \mathfrak{F}$.

Proof. Let μ be saturated, and let $A \subset X$ be such that $A \cap F \in \mathfrak{M}$ for each $F \in \mathfrak{F}$. Choose $B \in \mathfrak{M}$ with $\mu(B) < +\infty$. By (9.2) and (8.5) there are compact sets $F_1 \subset F_2 \subset \cdots \subset B$ such that $B - \bigcup_{n=1}^{\infty} F_n$ is a null set [see (8.7)]. Because

$$A \cap B = [A \cap (B - \bigcup_{n=1}^{\infty} F_n)] \cup \bigcup_{n=1}^{\infty} (A \cap F_n)$$

it follows from the completeness of μ that $A \cap B \in \mathfrak{M}$. Since μ is saturated, $A \in \mathfrak{M}$.

Since $\mu(F) < +\infty$ for each $F \in \mathfrak{F}$ [see (9.1)(ii)], the converse is trivial. ///

Since for regular measures the measure of a σ-finite set A can be approximated arbitrarily closely by the measures of compact subsets of A [see (9.3)], it is natural to ask, to what extent regular measures are determined by their values on compact sets. The next theorem and its corollary give a satisfactory answer to this question.

(9.9) Theorem. Let $\mathfrak{M}$ and $\mathfrak{N}$ be σ-algebras in X, and let μ and ν be regular measures on $\mathfrak{M}$ and $\mathfrak{N}$, respectively. Furthermore, let $\mu(F) = \nu(F)$ for each $F \in \mathfrak{F}$. If ν is complete and saturated, then $\mathfrak{M} \subset \mathfrak{N}$ and $\mu(A) = \nu(A)$ for each $A \in \mathfrak{M}$.

Proof. Assume that ν is complete and saturated.

(a) If $A \in \mathfrak{G}$, then by (9.1)(iii),

$$\begin{aligned}\mu(A) &= \sup\{\mu(F) : F \in \mathfrak{F},\ F \subset A\} \\ &= \sup\{\nu(F) : F \in \mathfrak{F},\ F \subset A\} = \nu(A)\end{aligned}$$

(b) Let $A \in \mathfrak{M}$ and $\mu(A) < +\infty$. Using (9.1)(iv), (9.2), (8.5) and (8.6) we can find sets $F_n \in \mathfrak{F}$ and $G_n \in \mathfrak{G}$ such that $\mu(G_1) < +\infty$,

$$F_n \subset F_{n+1} \subset A \subset G_{n+1} \subset G_n$$

$n = 1, 2, \ldots,$ and

$$\mu(A) = \mu\Big(\bigcup_{n=1}^{\infty} F_n\Big) = \mu\Big(\bigcap_{n=1}^{\infty} G_n\Big)$$

The sets $F = \bigcup_{n=1}^{\infty} F_n$ and $G = \bigcup_{n=1}^{\infty} G_n$ are Borel sets and so they belong to $\mathfrak{M} \cap \mathfrak{N}$. By (a), (8.5), and (8.6),

$$\nu(F) = \mu(F) \qquad \text{and} \qquad \nu(G) = \mu(G)$$

Thus

$$\nu(G - F) = \mu(G) - \mu(F) = 0$$

and since $A = F \cup (A - F)$, it follows from the completeness of ν that $A \in \mathfrak{N}$. Moreover,

$$\nu(A) = \nu(F) = \mu(F) = \mu(A)$$

(c) Let $A \in \mathfrak{M}$ be arbitrary and let $F \in \mathfrak{F}$. Then $A \cap F \in \mathfrak{M}$ and $\mu(A \cap F) < +\infty$ [see (9.1)(ii)]. Thus by (b), $A \cap F \in \mathfrak{N}$. Since ν is saturated, it follows from (9.8) that $A \in \mathfrak{N}$. Moreover, using (a) and (9.1)(iv) we obtain

$$\begin{aligned}\mu(A) &= \inf\{\mu(G) : G \in \mathfrak{G},\ A \subset G\} \\ &= \inf\{\nu(G) : G \in \mathfrak{G},\ A \subset G\} = \nu(A) \quad ///\end{aligned}$$

(9.10) Corollary. Let $\mathfrak{M}$ and $\mathfrak{N}$ be σ-algebras in X, and let μ and ν be regular, complete, saturated measures on $\mathfrak{M}$ and $\mathfrak{N}$, respectively. If $\mu(F) = \nu(F)$ for each $F \in \mathfrak{F}$, then $\mathfrak{M} = \mathfrak{N}$ and $\mu = \nu$.

Exercises

(9-1) Let X be a discrete space, and let $\aleph$ be an infinite cardinal. For $A \subset X$, set $\mu(A) = 0$ if $\|A\| \leq \aleph$, and $\mu(A) = +\infty$ otherwise. Show that

(i) μ is a measure on exp X.

(ii) μ is regular iff $\|X\| \leq \aleph$.

In exercises (9-2) through (9-9) we shall assume that X is a locally compact Hausdorff space and that $\mathfrak{M}$ is a σ-algebra in X which contains all open subsets of X.

(9-2)[+] Prove the following:

(i) The counting measure on $\mathfrak{M}$ [see (8-1)(iv)] is regular iff X is discrete.

(ii) Let X be discrete, and let μ be the weighted counting measure on $\mathfrak{M}$ determined by a function f on X [see (8-1)(vi)]. Then μ is regular iff f is real-valued.

(9-3) Let $\{x_n\} \subset X$, $\{a_n\} \subset [0, +\infty)$, and let $\Sigma_{n=1}^{\infty} a_n$ converge. For $A \in \mathfrak{M}$ set

$$\mu(A) = \sum_{n=1}^{\infty} a_n \chi_A(x_n)$$

and show that μ is a regular measure on $\mathfrak{M}$. Conclude that each Dirac measure on $\mathfrak{M}$ [see (8-1)(v)] is regular.

(9-4)[*+] Let $\mathfrak{B}$ be the Borel σ-algebra in X. A measure μ on $\mathfrak{B}$ is called a Borel measure whenever $\mu(F) < +\infty$ for every compact set $F \subset X$. Thus any regular measure on $\mathfrak{B}$ is a Borel measure, however, the converse is not always true [see (9-1) or (9-10)(vi)]. Let $A \subset X$ be a countable set, and let μ be a Borel measure such that $\mu(B) = \mu(B \cap A)$ for each

$B \in \mathfrak{B}$. Show that if X is paracompact, then μ is regular. Hint: Use the fact that every paracompact, locally compact Hausdorff space is a free union of σ-compact spaces (see Ref. 8, Chapter XI, Theorem 7.3, p. 241).

(9-5)+ Let μ be a measure on $\mathfrak{M}$, and let $\bar{\mu}$ on $\bar{\mathfrak{M}}$ be the completion of μ [see (8-9)]. Show that μ is regular iff $\bar{\mu}$ is.

(9-6)+ Let μ be a measure on $\mathfrak{M}$, and let $\mu^\wedge$ on $\mathfrak{M}$ be the saturation of μ [see (8-10)]. Show that

(i) μ is regular iff $\mu^\wedge$ is.

(ii) Suppose that μ is a regular measure. Let $\mathfrak{M}'$ be a σ-algebra in X containing $\mathfrak{M}$, and let μ' be a regular saturated measure on $\mathfrak{M}'$ such that $\mu'(A) = \mu(A)$ for each $A \in \mathfrak{M}$. Then $\mathfrak{M}^\wedge \subset \mathfrak{M}'$ and $\mu'(A) = \mu^\wedge(A)$ for each $A \in \mathfrak{M}$. Hint: If $B \in \mathfrak{M}'$ and $\mu'(B) < +\infty$, then there is an open set $G \subset X$ such that $B \subset G$ and $\mu(G) < +\infty$.

Thus among regular measures the saturation $\mu^\wedge$ is the unique minimal saturated extension of μ.

(9-7)+ Let μ be a measure on $\mathfrak{M}$, and let ν be the restriction of μ to the Borel σ-algebra in X. Show that

(i) If μ is regular, then ν is a regular Borel measure [see (9-4)].

(ii) If μ is regular, complete, and saturated, then μ is the completion and saturation (in this order) of ν. Hint: Use (i), (9-5), (9-6)(i), (8-10)(iii), and (9.10).

(iii) If μ is regular, complete, and σ-finite, then μ is the completion of ν. Hint: Use (i), (9-5), (8-9)(iv), (8.16), and (9.10).

A good example illustrating this exercise will be given in (11-3).

(9-8)*+ Let μ be a measure on $\mathfrak{M}$. A closed set $F \subset X$ is called a support of μ if $\mu(X - F) = 0$ and $\mu(G \cap F) > 0$ for every open set $G \subset X$ for which $G \cap F \neq \emptyset$. Show that

(i) The measure μ has at most one support. If it exists, it is denoted by supp μ.

(ii) If μ is regular, supp μ exists. Hint: Use (9.6) to find the complement of supp μ.

(iii) Let μ be a regular measure, and let $\bar{\mu}$, $\mu^\wedge$, and ν denote, respectively, the completion of μ, the saturation of μ, and the restriction of μ to the Borel σ-algebra in X. Then

$$\operatorname{supp} \mu = \operatorname{supp} \bar{\mu} = \operatorname{supp} \mu^\wedge = \operatorname{supp} \nu$$

(iv) Let μ be regular and σ-finite. If X is paracompact, then supp μ is σ-compact. Hint: Use the fact that every paracompact locally compact Hausdorff space is a free union of σ-compact spaces [see Ref. 8, Chapter XI, Theorem 7.3, p. 241] and show that the closure of a σ-compact subset of X is also σ-compact.

(v) Let μ be a regular measure which takes only finitely many real values. Then μ is a finite linear combination of Dirac measures. Hint: Observe that supp μ is a finite set.

(9-9)* Let X be σ-compact and first countable [see Ref. 16, Chapter 1, p. 50], and let μ be a measure on $\mathfrak{M}$ which has only finitely many real values. Prove that μ is regular. Hint: Show that μ is a finite linear combination of Dirac measures.

(9-10)+ The following is an important example of a finite nonregular measure.

(i) Let $X = W^-$ (see Chapter 1, Section B), and let $\mathfrak{C}$ be the system of all closed uncountable subsets of W. Observe that $A \subset W$ is uncountable iff it is cofinal (i.e., if for every $\beta \in W$, there is an $\alpha \in A$ such that $\beta < \alpha$).

(ii) Use (1-14) to show that $A \cap B \in \mathfrak{C}$ whenever $A \in \mathfrak{C}$ and $B \in \mathfrak{C}$.

(iii) Show that if $\{A_n\} \subset \mathfrak{C}$ is a decreasing sequence and $A = \bigcap_{n=1}^{\infty} A_n$, then $A \in \mathfrak{C}$. Hint: Given $\beta \in W$, choose $\alpha_n \in A_n$ so that $\beta < \alpha_1 < \alpha_2 < \cdots$, and observe that $\lim \alpha_n$ belongs to A.

(iv) Let $\mathfrak{M}$ consist of all $B \subset X$ for which there is an $A \in \mathfrak{C}$ such that either $A \subset B$ or $A \subset X - B$. Show that $\mathfrak{M}$ is a σ-algebra in X containing all Borel subsets of X. Hint: Show that $\mathfrak{M}$ contains all closed subsets of X.

(v) For $B \in \mathfrak{M}$ set $\mu(B) = 1$ if there is an $A \in \mathfrak{C}$ such that $A \subset B$, and $\mu(B) = 0$ otherwise. Prove that μ is a nonregular measure on $\mathfrak{M}$. Hint:

Show that $\mu(\{\Omega\}) = 0$, while $\mu(U) = 1$ for every neighborhood U of Ω.

(vi) Let ν be the restriction of μ to the Borel σ-algebra in X. Observe that ν is a nonregular Borel measure [see (9-4)].

(vii) Prove that the measures μ and ν have no support [see (9-8)].

(9-11)$^+$ Let X be a locally compact Hausdorff space, $\mathfrak{M}$ and $\mathfrak{N}$ be σ-algebras in X, and let μ and ν be regular, complete, saturated measures on $\mathfrak{M}$ and $\mathfrak{N}$, respectively. Show that if $\mu(C) = \nu(C)$ for each $C \in \mathfrak{F} \cap \mathfrak{G}_\delta$, then $\mathfrak{M} = \mathfrak{N}$ and $\mu = \nu$. Hint: If $F \in \mathfrak{F}$ and $G \in \mathfrak{G}$ with $F \subset G$, then there is a $C \in \mathfrak{F} \cap \mathfrak{G}_\delta$ such that $F \subset C \subset G$ [see (13.3)]. Use this and (9.1)(iv) to show that $\mu(F) = \nu(F)$ for each $F \in \mathfrak{F}$. Then apply (9.10).

Let X be an arbitrary Hausdorff space and let $\mathfrak{B}$ be the Borel σ-algebra in X. A measure μ on $\mathfrak{B}$ is called locally finite if each $x \in X$ has a neighborhood $U \in \mathfrak{B}$ with $\mu(U) < +\infty$. It is easy to see that each locally finite measure in X is a Borel measure in X [see (9-4)]. Moreover, if X is locally compact, the family of locally finite measures in X coincides with the family of Borel measures in X. In exercises (9-12) and (9-13) we shall give examples of Borel measures which are not locally finite.

(9-12) Let X be the set of all real numbers with the half-open interval topology [see Ref. 16, Chapter 1, problem K, p. 59]. For a Borel set $B \subset X$, let $\mu(B) = 0$ if B is countable, and let $\mu(B) = +\infty$ otherwise. Show that μ is a Borel measure in X which is not locally finite.

(9-13) Let X be the set of all irrational numbers with the usual topology. For a Borel set $B \subset X$, let $\mu(B) = 0$ if B is a subset of a σ-compact set, and let $\mu(B) = +\infty$ otherwise. Show that μ is a Borel measure in X which is not locally finite. Hint: Use the fact that X is not an F_σ subset of $\mathbf{R}$ [see Ref. 8, Chapter XI, Section 10, exercise 2, p. 249].

Let X be a locally compact Hausdorff space and let $\mathfrak{M}$ be a σ-algebra in X containing all open subsets of X. A regular measure μ on $\mathfrak{M}$ is called strongly regular if

$$\mu(A) = \sup\{\mu(F) : F \in \mathfrak{F},\ F \subset A\}$$

for each $A \in \mathfrak{M}$.

(9-14) Prove the following:

(i) Each σ-finite regular measure is strongly regular.

(ii) The completion [see (8-9)] of a strongly regular measure is strongly regular.

(iii) The counting measure [see (8-1)(iv)] in an uncountable discrete space is strongly regular but not σ-finite.

In (11-3) we shall see an example of a regular measure which is not strongly regular.

In exercises (9-15) through (9-18) we shall assume that X is a locally compact Hausdorff space, that $\mathfrak{M}$ is a σ-algebra in X, and that μ is a strongly regular measure on $\mathfrak{M}$.

(9-15) Let X be paracompact and let μ be diffused [see (8-11)]. Show that μ is σ-finite. Hint: Let X be a free union of σ-compact spaces X_α [see Ref. 8, Chapter XI, Theorem 7.3, p. 241]. Choose $x_\alpha \in X_\alpha \cap \text{supp}\ \mu$ whenever this intersection is nonempty, and denote by A the set consisting of these x_α's. Observe that A is closed and that $\mu(A) = 0$. Choose an open set G so that $A \subset G$ and $\mu(G) < +\infty$. Use (5-8) to prove that A is countable.

(9-16)* Prove that there is a disjoint family $\mathfrak{C}$ of nonempty compact sets with the following properties:

(i) If $C \in \mathfrak{C}$ and $C \cap G \neq \emptyset$ for an open set G, then $\mu(C \cap G) > 0$;

(ii) If $A \in \mathfrak{M}$ and $A \cap C = \emptyset$ for each $C \in \mathfrak{C}$, then $\mu(A) = 0$. Hint: Use Zorn's lemma to obtain a maximal family $\mathfrak{C}$ among all disjoint families of nonempty compact sets satisfying (i).

Note: The family $\mathfrak{C}$ from this exercise is called a concassage of μ.

(9-17)* Let $\mathfrak{C}$ be a concassage of μ and let $C_0 = \bigcup\{C : C \in \mathfrak{C}\}$. Prove the following:

(i) If $A \in \mathfrak{M}$ and $\mu(A) < +\infty$, then $\{C \in \mathfrak{C} : A \cap C \neq \emptyset\}$ is a countable family. Hint: Choose an open set G such that $A \subset G$ and $\mu(G) < +\infty$. Use (9-16)(i) and (5-8) to observe that $\{C \in \mathfrak{C} : G \cap C \neq \emptyset\}$ is a countable family.

(ii) If $A \in \mathfrak{M}$, then

$$\mu(A) = \Sigma\{\mu(A \cap C) : C \in \mathfrak{C}\}$$

Hint: Observe that by (i) and (5-9), the equation holds if A is compact. Apply the strong regularity of μ.

(iii) If μ is saturated, then $A \subset C_o$ is measurable iff $A \cap C$ is measurable for each $C \in \mathfrak{C}$. Hint: Use (i).

(iv) If μ is saturated, then $C_o \in \mathfrak{M}$ and $\mu(X - C_o) = 0$. Hint: Use (iii) and (9-16)(ii).

(v) If μ is saturated and complete, then $A \subset X$ is measurable iff $A \cap C$ is measurable for each $C \in \mathfrak{C}$. Hint: Use (iii) and (iv).

(9-18) Let μ be a diffused measure [see (8-11)]. Show that μ is σ-finite iff it is saturated. Hint: Let μ be saturated, and let $\mathfrak{C}$ be a concassage of μ. For each $C \in \mathfrak{C}$, choose $x_C \in C$ and let $A = \{x_C : C \in \mathfrak{C}\}$. It follows from (9-17) that $A \in \mathfrak{M}$, $\mu(A) = 0$, and consequently, that A is countable.

10. MEASURES INDUCED BY INTEGRALS

We shall use the results obtained in Chapters 2-6 to define measures by means of integrals. These measures, in turn, will provide additional information about the integrals which defined them.

The process by which an integral generates a measure is geometrically quite clear. Let A be a subset of a set X, and let $A^\wedge$ be the "cylinder" with base A and height 1. Intuitively speaking, an integral of the characteristic function of A is a "volume" of $A^\wedge$ (see Fig. 10.1). Since the height of A is 1, numerically, the "volume" of $A^\wedge$ equals to the "area" of A. This "area" will be called the measure of A induced by the integral.

Therefore we shall call a set $A \subset X$ measurable whenever its characteristic function χ_A is integrable, i.e., whenever $\chi_A \in \mathcal{L}_+$. Obviously, the measurable sets should form a σ-algebra in X. In particular, the set X should be measurable. Since the function χ_X is identically equal to 1, X will be measurable iff $1 \wedge f \in \mathcal{L}$ for each $f \in \mathcal{L}$ [see (6.9)]. This motivates the following definition introduced by M. H. Stone [22].

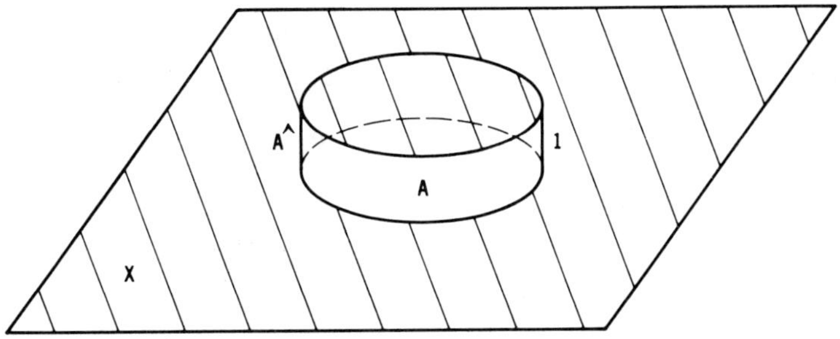

FIG. 10.1

(10.1) Definition. A system $\mathcal{E}$ of functions on a set X is called a Stonian system whenever

$$f \in \mathcal{E} \Rightarrow 1 \wedge f \in \mathcal{E}$$

Unless specified otherwise, we shall assume throughout this chapter that X is an arbitrary set, $\mathfrak{F}$ is a fundamental system on X, and that I is an integral on $\mathfrak{F}$ which has been extended to $\bar{\mathfrak{L}}$.

We begin with an immediate but useful consequence of (6.9).

(10.2) If $\mathfrak{F}$ is a Stonian system, so is $\mathfrak{L}$.

The converse of this statement is incorrect [see (11-4)]. We note that with the exception of exercises (2-8)(b), (5-6), and (11-4), all fundamental systems considered in this text are Stonian. In particular, if X is a topological space, the systems $C(X)$ and $C_o(X)$ are Stonian.

Throughout this chapter we shall assume that $\mathfrak{L}$ is a Stonian system.

(10.3) Theorem. Let $\mathfrak{I}$ consist of all sets $A \subset X$ with $\chi_A \in \mathfrak{L}_+$, and let $\iota(A) = I\chi_A$ for each $A \in \mathfrak{I}$. Then $(X, \mathfrak{I}, \iota)$ is a measure space.

Proof. (a) Since $\mathfrak{L}$ is a Stonian system, $X \in \mathfrak{I}$.

(b) Let $A \in \mathfrak{I}$. If $f \in \mathfrak{F}$ and $f \geq 0$, then

$$\chi_{X-A} \wedge f = \chi_X \wedge f - \chi_A \wedge f$$

belongs to $\mathfrak{L}$. By (6.7), $X - A \in \mathfrak{I}$.

(c) Let $\{A_n\} \subset \mathfrak{I}$ be a countable family, and let $A = \bigcup_n A_n$. Adding empty sets if necessary, we may assume that $\{A_n\}$ is an infinite sequence. If $f_n = \vee_{i=1}^n \chi_{A_i}$, then $f_n \in \mathfrak{L}_+$, $n = 1, 2, \ldots$ [see (6.3)], and $f_n \nearrow \chi_A$. It follows from (6.8) that $A \in \mathfrak{I}$.

(d) By (4.2), $\iota(\emptyset) = 0$.

(e) Let $\{A_n\} \subset \mathfrak{I}$ be a countable disjoint family, and let $A = \bigcup_n A_n$. Then $\chi_A = \Sigma_n \chi_{A_n}$ [see (7.23)], and so by (6.15) or (6.18),

$$\iota(A) = \sum_n \iota(A_n) \quad ///$$

(10.4) Definition. The measure space $(X, \mathfrak{J}, \iota)$ from Theorem (10.3) is said to be induced by the integral I.

Throughout this chapter the words measure and measurability will always refer to the measure space $(X, \mathfrak{J}, \iota)$ induced by the integral I.

We shall denote by $\mathcal{J}$ the family of all measurable functions on X, and we shall find its relation to the family $\bar{\mathcal{L}}$ of all integrable functions on X.

(10.5) Lemma. Let $f \in \mathcal{J}$ and $f \geq 0$. Then $f \in \mathcal{L}_+$.

Proof. By (7.29) there is a sequence $\{f_n\}$ of nonnegative $\mathfrak{J}$-simple functions on X such that $f_n \nearrow f$. It follows from (7.26), (6.14), and (6.15) that $f_n \in \mathcal{L}_+$, n = 1, 2, ..., and so by (6.8) also $f \in \mathcal{L}_+$. ///

From (7.16) we obtain the following:

(10.6) Corollary. If $f \in \mathcal{J}$, then $f^+ \in \mathcal{L}_+$ and $f^- \in \mathcal{L}_+$.

(10.7) Proposition. Let $f \in \mathcal{L}$, and let $0 < a < +\infty$. If

$$A = \{x \in X : f(x) > a\}$$

then $A \in \mathfrak{J}$ and $\iota(A) < +\infty$.

Proof. Because $\mathcal{L}$ is a Stonian system, it follows from (6.14) that $a \in \mathcal{L}_+$. Thus f - a belongs to $\mathcal{L}_-$ [see (6.2) and the dual of (6.15)]. By (5.14) and the dual of (6.4), the functions

$$f_n = [n(f - a)^+] \wedge 1$$

n = 1, 2, ..., are summable. It is easy to see that $f_n \nearrow \chi_A$. Hence, $A \in \mathfrak{J}$ by (6.8). Since $\chi_A \leq f^+/a$, we have

$$\iota(A) = I\chi_A \leq \frac{1}{a} If^+ < +\infty$$

for by (5.22), $f^+ \in \mathcal{L}$. ///

(10.8) Corollary. If $f \in \mathcal{L}$, then the set

$$A = \{x \in X : f(x) \neq 0\}$$

is measurable and σ-finite.

Proof. Because

$$A = \bigcup_{n=1}^{\infty} \{x \in X : |f(x)| > \frac{1}{n}\}$$

the corollary follows from (5.19) and (10.7). ///

(10.9) Proposition. Let $f \in \mathfrak{L}$ and $A \in \mathfrak{J}$. Then $f\chi_A \in \mathfrak{L}$.

Proof. By (6.9) and its dual, the functions

$$f_n = (f \wedge n\chi_A) \vee (-n\chi_A)$$

$n = 1, 2, \ldots$, are summable. Since $-f^- \leq f_n \leq f^+$ and $f_n \to f\chi_A$, the proposition follows from (5.17). ///

(10.10) Corollary. Let $f \in \mathfrak{L}_+$ and $A \in \mathfrak{J}$. Then $f\chi_A \in \mathfrak{L}_+$.

Proof. Let $f \geq 0$, and choose $g \in \mathfrak{F}$, $g \geq 0$. Since

$$(f\chi_A) \wedge g = (f \wedge g)\chi_A$$

it follows from (10.9) and (6.7) that $f\chi_A \in \mathfrak{L}_+$.

If $f \in \mathfrak{L}_+$ is arbitrary, it follows from the first part of the proof (6.4), and (10.9) that $f^+\chi_A \in \mathfrak{L}_+$ and $f^-\chi_A \in \mathfrak{L}$. Thus according to (6.15),

$$f\chi_A = f^+\chi_A - f^-\chi_A$$

belongs to $\mathfrak{L}_+$. ///

(10.11) Lemma. Let f be a nonnegative function on X, and let $f \wedge g \in \mathfrak{L}$ for each nonnegative $g \in \mathfrak{L}$ for which

$$\iota(\{x \in X : g(x) > 0\}) < +\infty$$

Then $f \in \mathfrak{L}_+$.

Proof. Choose $g \in \mathfrak{F}$, $g \geq 0$, and let

$$A_n = \{x \in X : g(x) > \frac{1}{n}\}$$

$n = 1, 2, \ldots$. Setting $g_n = g\chi_{A_n}$, we have

$$\{x \in X : g_n(x) > 0\} = A_n$$

and $g_n \nearrow g$; for $A_n \subset A_{n+1}$ and

$$\bigcup_{n=1}^{\infty} A_n = \{x \in X : g(x) > 0\}$$

By (10.7) and (10.9), $g_n \in \mathfrak{L}$ and $\iota(A_n) < +\infty$, $n = 1, 2, \ldots$. According to our assumption, $f \wedge g_n \in \mathfrak{L}$. Because $0 \leq f \wedge g_n \leq g$ and $f \wedge g_n \nearrow f \wedge g$, we have $f \wedge g \in \mathfrak{L}$ [see (5.17)]. The lemma follows from (6.7). ///

(10.12) Lemma. Let $f \in \mathfrak{L}_+$ and $a \in \mathbf{R}$. Then $(f + a)^+ \in \mathfrak{L}_+$.

Proof. Let $f \leq b$ for some $b \in \mathbf{R}$. Choose a nonnegative $g \in \mathfrak{L}$ so that the set

$$A = \{x \in X : g(x) > 0\}$$

has finite measure. We have $f\chi_A \in \mathfrak{L}_+$ [see (10.8) and (10.10)] and

$$I(f\chi_A) \leq bI\chi_A = b\iota(A) < +\infty$$

Thus according to (6.12) both functions χ_A and $f\chi_A$ are summable. Because

$$(f + a)^+ \wedge g = [(f + a)^+\chi_A] \wedge g = (f\chi_A + a\chi_A)^+ \wedge g$$

it follows from (10.11) that $(f + a)^+ \in \mathfrak{L}_+$.

If $f \in \mathfrak{L}_+$ is arbitrary, let $f_n = f \wedge n$, $n = 1, 2, \ldots$. Since $(f_n + a)^+ \nearrow (f + a)^+$, the lemma follows from the first part of the proof and (6.8). ///

(10.13) Corollary. $\bar{\mathfrak{L}} \subset \mathfrak{I}$.

Proof. Let $f \in \mathfrak{L}_+$, $a \in \mathbf{R}$, and let

$$A = \{x \in X : f(x) > a\}$$

By (10.12), the functions

$$f_n = [n(f - a)^+] \wedge 1$$

$n = 1, 2, \ldots,$ belong to $\mathcal{L}_+$. Since $f_n \nearrow \chi_A$, it follows from (6.8) that $A \in \mathfrak{I}$. Thus $f \in \mathcal{J}$ according to (7.9).

It follows from the first part of the proof, (6.2), and (7.14) that also $\mathcal{L}_- \subset \mathcal{J}$. ///

Combining Corollaries (10.6) and (10.13) we obtain the following important result.

(10.14) Theorem. Let f be a function on X. Then $f \in \mathcal{J}$ iff $f^+ \in \mathcal{L}_+$ and $f^- \in \mathcal{L}_+$.

(10.15) Corollary. Let f be a function on X. Then $f \in \bar{\mathcal{L}}$ iff $f \in \mathcal{J}$ and $\bar{I}f^+ - \bar{I}f^-$ has meaning.

Proof. The corollary follows easily from (6.4), (10.14), (6.12) and (6.15). ///

Corollary (10.15) is extremely useful as it reduces many problems of integration theory to the study of upper integrals of nonnegative measurable functions.

(10.16) Corollary. Let f be a function on X. Then $f \in \mathcal{L}$ iff $f \in \mathcal{J}$ and $|f| \in \mathcal{L}$.

Proof. Let $f \in \mathcal{J}$ and $|f| \in \mathcal{L}$. Since $f^+ \leq |f|$ and $f^- \leq |f|$, it follows from (10.14) and (6.12) that $f \in \mathcal{L}$.

The converse follows from (10.13) and (5.19). ///

(10.17) Corollary. Let $f \in \mathcal{L}$, $g \in \mathcal{J}$, and let $|g| \leq a$ for some $a \in \mathbf{R}$. Then $fg \in \mathcal{L}$.

Proof. By (10.13) and (7.18), $fg \in \mathcal{J}$ and

$$I(|fg|) \leq aI(|f|) < +\infty$$

The corollary follows from (10.16). ///

Before closing this chapter we shall establish some additional properties of the induced measure ι.

(10.18) Theorem. The measure ι is complete and saturated.

Proof. Let $A \in \mathfrak{I}$, $\iota(A) = 0$, and let $B \subset A$. Then

$$0 \le \underline{I}\chi_B \le \bar{I}\chi_B \le I\chi_A = \iota(A) = 0$$

and it follows that $B \in \mathfrak{I}$.

Let $A \subset X$, and let $A \cap B \in \mathfrak{I}$ for each $B \in \mathfrak{I}$ for which $\iota(B) < +\infty$. Choose a nonnegative $g \in \mathfrak{L}$ so that the set

$$C = \{x \in X : g(x) > 0\}$$

has a finite measure. By (10.13) and (7.29), there is a sequence $\{g_n\}$ of nonnegative $\mathfrak{I}$-simple functions on X such that $g_n \nearrow g$. If the expression

$$g_n = \sum_{i=1}^{k_n} b_{i,n}\chi_{B_{i,n}}$$

is the canonical form of g_n [see (7.27)], it is easy to see that

$$g_n \wedge \chi_A = \sum_{i=1}^{k_n} (b_{i,n} \wedge 1)\chi_{B_{i,n} \cap A}$$

Since $\iota(C) < +\infty$, we have $\iota(B_{i,n}) < +\infty$ whenever $b_{i,n} > 0$. Thus according to our assumption, $\chi_{B_{i,n} \cap A} \in \mathfrak{L}$ whenever $b_{i,n} > 0$. Consequently, $g_n \wedge \chi_A \in \mathfrak{L}$. By (5.17) also $g \wedge \chi_A \in \mathfrak{L}$; for $0 \le g_n \wedge \chi_A \le g$. It follows from (10.11) that $A \in \mathfrak{I}$. ///

(10.19) Theorem. Let X be a locally compact Hausdorff space, $\mathfrak{F} = C_o(X)$, and let I be a Bourbaki integral on $\mathfrak{F}$. Then the induced measure ι is regular.

Proof. Denote by $\mathfrak{Z}$ and $\mathfrak{G}$ the families of all compact and open subsets of X, respectively.

(a) Let $G \in \mathfrak{G}$. Then $G \in \mathfrak{I}$, for it follows from (3.5) that $\chi_G \in \mathfrak{F}_+$.

(b) Let $F \in \mathfrak{Z}$. By the local compactness of X, there is a $G \in \mathfrak{G}$ for which $F \subset G$ and $G^- \in \mathfrak{Z}$. We can find a function $f \in C(X)$ so that $0 \le f \le 1$ and

$$f(x) = \begin{cases} 1 & \text{if } x \in F \\ 0 & \text{if } x \in X - G \end{cases}$$

Clearly, $\chi'_F \le f$, and since G^- is compact, $f \in \mathfrak{F}$. Thus

$$\iota(F) = I\chi_F \le If < +\infty$$

(c) Let $G \in \mathfrak{G}$ and let $\alpha < \iota(G)$. Since $G \in \mathfrak{F}_+$, there is an $f \in \mathfrak{F}$ such that $f \le \chi_G$ and $If > \alpha$. The sets

$$F_n = \{x \in X : f(x) \ge \frac{1}{n}\}$$

$n = 1, 2, \ldots$, are compact subsets of G and $F_n \nearrow F$, where $F = \bigcup_{n=1}^{\infty} F_n$. Clearly, $f \le \chi_F$, and by (8.5),

$$\alpha < If \le I\chi_F = \iota(F) = \lim \iota(F_n) \le \sup\{\iota(F) : F \in \mathfrak{J}, F \subset G\}$$

It follows from the arbitrariness of α and (8.3) that

$$\iota(G) \le \sup\{\iota(F) : F \in \mathfrak{J}, F \subset G\} \le \iota(G)$$

(d) Let $A \in \mathfrak{J}$. By (8.3),

$$\iota(A) \le \inf\{\iota(G) : G \in \mathfrak{J}, A \subset G\}$$

and the equality holds trivially if $\iota(A) = +\infty$. Thus assume that $\iota(A) < +\infty$ and choose $\epsilon \in (0,1)$. There is an $f \in \mathfrak{F}_+$ such that

$$f \ge \chi_A \quad \text{and} \quad If < I\chi_A + \epsilon$$

The set $G = \{x \in X : f(x) > 1 - \epsilon\}$ is open [see (3.5)], $A \subset G$, and $\chi_G \le f/1 - \epsilon$. Thus

$$\iota(G) = I\chi_G \le \frac{1}{1-\epsilon} If < \frac{1}{1-\epsilon}(I\chi_A + \epsilon) = \frac{1}{1-\epsilon}[\iota(A) + \epsilon]$$

From the arbitrariness of ϵ it follows that

$$\iota(A) \le \inf\{\iota(G) : G \in \mathfrak{J}, A \subset G\} \le \iota(A) \quad ///$$

Exercises

(10-1) Let X be a discrete space, $x_o \in X$, and let I_1 and I_2 denote the Dirac integral at x_o [see (4-1)] viewed, respectively, as a Daniell integral and a Bourbaki integral on $\mathfrak{F} = C_o(X)$. Let $(X, \mathfrak{I}_i, \iota_i)$ be the measure space induced by I_i, $i = 1, 2$. Show that

(i) $\mathfrak{I}_1 = \mathfrak{I}_2 = \exp X$.

(ii) ι_2 is the Dirac measure at x_o [see (8-1)(v)].

(iii) ι_1 is the Dirac measure at x_o iff X is countable.

(iv) If X is uncountable, then ι_1 is not regular.

(10-2) Let X be a discrete space, and let $\mathfrak{F} = C_o(X)$. Setting $If = \Sigma_{x \in X} f(x)$ for $f \in \mathfrak{F}$, we have defined a Bourbaki integral I on $\mathfrak{F}$ [see (4-4)]. Show that

(i) If $(X, \mathfrak{I}, \iota)$ is the measure space induced by I, then $\mathfrak{I} = \exp X$ and ι is the counting measure [see (8-1)(iv)].

(ii) The result from (i) remains unchanged if we view I as a Daniell integral on $\mathfrak{F}$.

(10-3) Let X be an uncountable set, and let $(X, \mathfrak{I}, \iota)$ be the measure space induced by the integral from exercise (4-2). Show that

(i) $\mathfrak{I}$ consists of all countable subsets of X and their complements.

(ii) For $A \in \mathfrak{I}$, $\iota(A) = 0$ if A is countable, and $\iota(A) = 1$ otherwise.

(10-4) Let $\aleph$ be an infinite cardinal and $\|X\| > \aleph$. Show that

(i) The integral from exercise (4-3) induces the measure space $(X, \mathfrak{M}, \mu)$ from exercise (8-2).

(ii) The measure space $(X, \mathfrak{M}, \nu)$ from exercise (8-2) is not induced by any integral.

(10-5)$^+$ Let $X = W^-$ [see Chapter 1, Section B], and denote by I_1 and I_2 the Dirac integral at Ω [see (4-1)] viewed, respectively, as a Daniell integral and a Bourbaki integral on $\mathfrak{F} = C(X)$. Let $(X, \mathfrak{I}_i, \iota_i)$ be the measure space induced by I_i, $i = 1, 2$. Show that

(i) $\mathfrak{I}_1$ consists of all countable subsets of W and their complements in X.

(ii) $\mathfrak{I}_2 = \exp X$.

(iii) ι_i, $i = 1, 2$, is the Dirac measure at Ω [see (8-1)(v)].

(10-6) Let X be a locally compact Hausdorff space, $\mathfrak{F} = C_o(X)$, and let I be a Bourbaki integral on $\mathfrak{F}$. If $(X, \mathfrak{I}, \iota)$ is the measure space induced by I, show that supp ι = supp I [see (4-14) and (9-8)].

(10-7)$^+$ Let $\mathfrak{F} = C_o(\mathbf{R}^m)$, $m \geq 1$ an integer, and let I be the integral on $\mathfrak{F}$ defined in (4-10). Denote by $(\mathbf{R}^m, \Lambda_m, \lambda_m)$ the measure induced by I and prove

(i) If $K = \Pi_{i=1}^m [a_i, b_i]$ and $K^o = \Pi_{i=1}^m (a_i, b_i)$, $a_i, b_i \in \mathbf{R}$, $a_i \leq b_i$, $i = 1, \ldots, m$, then

$$\lambda_m(K) = \lambda_m(K^o) = \Pi_{i=1}^m (b_i - a_i)$$

Hint: Approximate χ_K from above by functions from $\mathfrak{F}$ and calculate $\lambda_m(K)$. Observe that the faces of K are λ_m-null sets.

(ii) For an integer k, $0 \leq k < m$, $\mathbf{R}^k$ is a λ_m-null subset of $\mathbf{R}^m$.

The measure λ_m is extremely important. It is called the m-dimensional Lebesgue measure. If $m = 1$, we shall write Λ and λ instead of Λ_1 and λ_1, respectively, and we shall call the measure λ simply the Lebesgue measure.

(10-8)$^{*+}$ We shall show that $\Lambda \neq \exp \mathbf{R}$, i.e., there exists a Lebesgue nonmeasurable set.

(i) If $A \subset \mathbf{R}$ and $x_o \in \mathbf{R}$, let $x_o + A = \{x_o + x : x \in A\}$ be the translation of A by x_o. Show that λ is translation invariant, i.e., if $A \in \Lambda$ and $x_o \in \mathbf{R}$, then $x_o + A \in \Lambda$ and $\lambda(x_o + A) = \lambda(A)$. Hint: $\int_a^b f(x)\,dx = \int_{a+x_o}^{b+x_o} f(x - x_o)\,dx$ for each $f \in C_o(\mathbf{R})$ and $a, b, x_o \in \mathbf{R}$.

(ii) Let Q denote the set of all rational numbers. For $x, y \in \mathbf{R}$ write $x \sim y$ iff $x - y$ belongs to Q. Show that $\sim$ is an equivalence relation on $\mathbf{R}$ and that $x + Q$ is the equivalence class determined by $x \in \mathbf{R}$.

(iii) Use the axiom of choice to construct a set $M \subset \mathbf{R}$ such that $M \cap (x + Q)$ is a singleton for each $x \in \mathbf{R}$. Show that

$$\mathbf{R} = \bigcup\{x + M : x \in Q\}$$

and that this union is disjoint.

(iv) Assume that $M \in \Lambda$, and use (i) and (iii) to show that $\lambda(M) > 0$.

(v) Assume that $M \in \Lambda$, and show that there is a set $N \subset M$ and an integer $n \geq 1$ such that $N \in \Lambda$, $\lambda(N) > 0$, and $N \subset (-n,n)$. Hint: Use (iv) and (8.5).

(vi) Show that $P = \bigcup\{x + N : x \in Q \cap (0,1)\}$ is a disjoint countable union and that $P \subset (-n, n + 1)$.

(vii) Assume that $M \in \Lambda$, and use (v), (vi), and (i) to show that $P \in \Lambda$ and $\lambda(P) = +\infty$.

(viii) Observe that if $P \in \Lambda$, then by (vi) and (10-7)(i), $\lambda(P) \leq 2n + 1$.

Remark: The following observations about the existence of a Lebesgue nonmeasurable set are noteworthy.

(a) Our proof was based on translation invariance of the Lebesgue measure. It generalizes to other translation invariant measures.

(b) A different proof, which uses the regularity of the Lebesgue measure, can be given by application of exercise (10-14).

(c) The existence of a Lebesgue nonmeasurable set cannot be proved without the axiom of choice. This means that all subsets of **R** which can be defined in some "constructive" way are, in fact, Lebesgue measurable.

(10-9)$^+$ Let $\mathfrak{F} = C_o(\mathbf{R})$, let g be a nondecreasing real-valued function on **R**, and let I be the integral on $\mathfrak{F}$ defined in (4-7). Denote by $(\mathbf{R}, \Lambda_g, \lambda_g)$ the measure space induced by I, and show that

(i) $\lambda_g([a,b]) = g(b+) - g(a-)$ if $a,b \in \mathbf{R}$ and $a \leq b$.

(ii) $\lambda_g((a,b)) = g(b-) - g(a+)$ if $a,b \in \mathbf{R}^-$ and $a < b$.

(iii) If $g(x) = x$ for all $x \in \mathbf{R}$, then λ_g is the Lebesgue measure.

(iv) If $x_o \in \mathbf{R}$, $g(x) = 0$ for $x \in (-\infty, x_o)$ and $g(x) = 1$ for $x \in [x_o, +\infty)$, then $\Lambda_g = \exp \mathbf{R}$ and λ_g is the Dirac measure at x_o [see (8-1)(v)].

The measure λ_g is called the Lebesgue-Stieltjes measure induced by g.

(10-10) Let Λ be the family of all Lebesgue measurable subsets of $\mathbf{R}$, and let λ be the Lebesgue measure on Λ. Denote by $\mathfrak{M}$ the family of all Lebesgue measurable subsets of $(-\infty, 0)$ and their complements in $\mathbf{R}$, and let μ be the restriction of λ to $\mathfrak{M}$. If $\hat{\mu}$ on $\hat{\mathfrak{M}}$ is the saturation of μ [see (8-10)], show that

(i) $\hat{\mathfrak{M}} \not\subset \Lambda$. Hint: Use (10-8).

(ii) $[0,1] \subset \Lambda \cap \hat{\mathfrak{M}}$, and yet $\lambda([0,1]) \neq \hat{\mu}([0,1])$.

(10-11) For $f \in C_0(\mathbf{R})$, set

$$If = \lim_{a \to +\infty} \int_{-a}^{a} f(x)\, dx$$

By (10-9)(iii), I induces the Lebesgue measure λ. Using (10-8), prove that there is a Lebesgue nonmeasurable set $M \subset [0,1]$. Let

$$f = \chi_M + (+\infty)\chi_{[2,3]}$$

and let $g \in C_0(\mathbf{R})$ be such that $\chi_{[0,3]} \leq g$. Show that $\underline{I}f = \bar{I}f = +\infty$ and $\underline{I}(f \wedge g) < \bar{I}(f \wedge g)$.

(10-12)[+] Let X be an uncountable first countable compact Hausdorff space. Show that

(i) There are disjoint closed sets $A_i \subset X$ such that $\|A_i\| = \|X\|$, $i = 0, 1$. Hint: Show first that there is an $x \in X$ such that $\|U\| = \|X\|$ for each neighborhood U of x (such a point x is called a condensation point of X and it exists in any compact space X). Then use first countability at x.

(ii) There are closed nonempty sets $A_{i_1 \cdots i_n} \subset X$, $i_j = 0, 1$, $j = 1, \ldots, n$, $n = 1, 2, \ldots$, such that $A_{i_1 \cdots i_n i_{n+1}} \subset A_{i_1 \cdots i_n}$ and $A_{i_1 \cdots i_n 0} \cap A_{i_1 \cdots i_n 1} = \emptyset$. Hint: Iterate (i).

(iii) $\|X\| \geq c$. Hint: If $\{i_n\}$ is a sequence of zeros and ones, then use the axiom of choice to select a point $x_{i_1 i_2 \cdots} \in \bigcap_{n=1}^{\infty} A_{i_1 \cdots i_n}$. Show that the map $\{i_n\} \mapsto x_{i_1 i_2 \cdots}$ is one-to-one.

Note: For many years it was an outstanding question whether $\|X\| = c$. An affirmative answer was given in 1969 by A. V. Arhangelskij [2].

(10-13)* Let X be a second countable locally compact Hausdorff space and let μ be a diffused [see (8-11)] regular measure on a σ-algebra $\mathfrak{M} \subset \exp X$ such that $\mu(X) > 0$. We shall show that $\mathfrak{M} \neq \exp X$.

(i) Observe that X is first countable and σ-compact.

(ii) If $\mathfrak{C}$ is a family of all uncountable compact subsets of X, then $\|\mathfrak{C}\| \leq c$. Hint: Use (7-10).

(iii) There is an ordinal $\gamma \leq \zeta$ (see Chapter 1, Section A) and a one-to-one map $\alpha \mapsto F_\alpha$ from the set $\{\alpha : \alpha < \gamma\}$ onto $\mathfrak{C}$.

(iv) By transfinite induction, construct a set $\{x_\alpha, y_\alpha : \alpha < \gamma\}$ as follows: Choose distinct points x_0, $y_0 \in F_0$ and suppose that x_α, $y_\alpha \in F_\alpha$ have been already chosen for each ordinal α, $\alpha < \beta < \gamma$, so that

$$\{x_\alpha : \alpha < \beta\} \cap \{y_\alpha : \alpha < \beta\} = \emptyset$$

Let $B = \{x_\alpha, y_\alpha : \alpha < \beta\}$. Since $\|B\| < c$ and $\|F_\beta\| = c$ [see (i) and (10-12)(iii)], there are distinct points $x_\beta, y_\beta \in F_\beta - B$.

(v) Set $A = \{x_\alpha : \alpha < \gamma\}$, and observe that $F \cap A \neq \emptyset$ and $F - A \neq \emptyset$ for each $F \in \mathfrak{C}$.

(vi) Show that $A \notin \mathfrak{M}$. Hint: Suppose that $A \in \mathfrak{M}$. By (v) all compact subsets of A and X - A are countable. It follows from (i) and (9.3) that $\mu(A) = \mu(X - A) = 0$; a contradiction.

We note that a considerable generalization of this exercise can be found in Ref. 24.

(10-14) Denote by Λ_n', $n \geq 1$ an integer, the system of all sets $A \subset \mathbf{R}^n$ for which there is a set $B \in \Lambda_n$ such that $\|(A - B) \cup (B - A)\| < c$. If $A \in \Lambda_n'$, find a $B \in \Lambda_n$ so that $\|(A - B) \cup (B - A)\| < c$ and let $\lambda_n'(A) = \lambda_n(B)$. Show that

(i) Λ_n' is a σ-algebra in $\mathbf{R}^n$ containing Λ_n. Hint: Use (1-10).

(ii) λ_n' is a well-defined measure on Λ_n'. Hint: Use (10.19), (9.3) and (10-12)(iii) to show that if $A \in \Lambda_n$ and $\|A\| < c$, then $\lambda_n(A) = 0$.

(iii) The measure λ_n' extends λ_n and it is complete and σ-finite.

Note: This extension λ_n' of the n-dimensional Lebesgue measure is due to J. von Neumann. It is not clear whether $\Lambda_n = \Lambda_n'$ or $\Lambda_n \subsetneqq \Lambda_n'$. For instance, if the continuum hypothesis (see Chapter 1, Section A) is assumed, then $\Lambda_n' = \Lambda_n$; for under the continuum hypothesis if $\|A\| < c$, then A is countable.

11. ALMOST EVERYWHERE

If the integral of a nonnegative integrable function is zero, it does not necessarily follow that the function is zero. It does imply, however, that the function is zero with the exception of a "small" set. On the other hand, altering an integrable function on a "small" set does not change its integral. Our next goal is to give a precise formulation of these phenomena.

Throughout this chapter we shall assume the following:

(i) X is an arbitrary set

(ii) $\mathfrak{F}$ is a fundamental system on X

(iii) I is an integral on $\mathfrak{F}$ which has been extended to $\bar{\mathfrak{L}}$ and for which $\mathfrak{L}$ is a Stonian system

(iv) $(X, \mathfrak{J}, \iota)$ is the measure space induced by I and $\mathfrak{J}_o = \{A \in \mathfrak{J} : \iota(A) = 0\}$.

Recall that

$$\{A_n\}_{n=1}^{\infty} \subset \mathfrak{J}_o \Rightarrow \bigcup_{n=1}^{\infty} A_n \in \mathfrak{J}_o$$

and since the measure ι is complete [see (10.18)], also

$$(A \in \mathfrak{J}_o, \quad B \subset X) \Rightarrow A \cap B \in \mathfrak{J}_o$$

(11.1) Definition. Let P be a property of $x \in X$. We say that P holds almost everywhere whenever the set

$$X - \{x \in X : P(x)\}$$

belongs to $\mathfrak{J}_o$.

In other words, property P holds almost everywhere if the set of those $x \in X$ for which P(x) does not hold has measure zero. Sometimes, instead of saying that property P holds almost everywhere, we shall say that P(x) holds for almost all $x \in X$.

In particular, if f and g are functions on X, we say that f is equal to g almost everywhere and write $f \doteq g$, whenever

$$\{x \in X : f(x) \neq g(x)\} \in \mathfrak{J}_o$$

Similarly, we shall use symbols $f \dot{\leq} g$, $f \dot{<} g$, and $f_n \dot{\to} f$ to denote, respectively, that $f \leq g$ almost everywhere, $f < g$ almost everywhere, and that the sequence $\{f_n\}$ of functions on X converges to f almost everywhere.

(11.2) Proposition. Let f and g be functions on X. If $f \doteq g$, then

$$\bar{I}f = \bar{I}g \qquad \text{and} \qquad \underline{I}f = \underline{I}g$$

Proof. Let $A = \{x \in X : f(x) \neq g(x)\}$ and $h = (+\infty)\chi_A$. Since $\chi_A \in \mathfrak{L}$,

$$I(n\chi_A) = n\iota(A) = 0$$

$n = 1, 2, \ldots$, and $n\chi_A \nearrow h$, we have $h \in \mathfrak{L}$ and $Ih = 0$. Because

$$f(x) \leq g(x) + h(x)$$

whenever the right side has meaning, $\bar{I}f \leq \bar{I}g$ by (5.6). From the symmetry we obtain $\bar{I}g \leq \bar{I}f$. To complete the proof it suffices to apply (5.2). ///

Using this proposition, we can strengthen many of our previous results. We shall give an example.

(11.3) Theorem. Let f and f_n be functions on X such that $f_n \dot{\leq} f_{n+1}$, $n = 1, 2, \ldots$, and $f_n \dot{\to} f$. If $\bar{I}f_1 > -\infty$, then $\bar{I}f_n \nearrow \bar{I}f$.

Proof.

$$A_n = \{x \in X : f_n(x) \leq f_{n+1}(x)\} \qquad (n = 1, 2, \ldots)$$

$$A_o = \{x \in X : f_n(x) \to f(x)\}$$

By our assumption $\{X - A_n\}_{n=o}^{\infty} \subset \mathfrak{I}_o$, and so if $A = \bigcup_{n=o}^{\infty} (X - A_n)$, then $A \in \mathfrak{I}_o$. Let $g(x) = g_n(x) = 0$ if $x \in A$, and $g(x) = f(x)$ and $g_n(x) = f_n(x)$ if $x \in X - A$, $n = 1, 2, \ldots$. Then $g \doteq f$, $g_n \doteq f_n$, and $g_n \nearrow g$. The theorem follows from (5.9) and (11.2). ///

We recommend that the reader go through all assertions from Chapters 5 and 6 and try to generalize them by using almost everywhere assumptions whenever possible.

(11.4) Proposition. Let f be a function on X. If $\bar{I}f < +\infty$, then $f \dot{<} +\infty$.

Proof. Let $\bar{I}f < +\infty$. Then there is a $g \in \mathfrak{F}_+$ such that $g \geq f$ and $Ig < +\infty$. Thus g belongs to $\mathfrak{L}$ and so does g^+. Since $f^+ \leq g^+$, we have $\bar{I}f^+ \leq Ig^+ < +\infty$. If $A = \{x \in X : f(x) = +\infty\}$, then $\chi_A \leq f^+/n$, $n = 1, 2, \ldots$. Therefore,

$$0 \leq \underline{I}\chi_A \leq \bar{I}\chi_A \leq \frac{1}{n}\bar{I}f^+$$

$n = 1, 2, \ldots$, and it follows that $A \in \mathfrak{I}_o$. ///

(11.5) Corollary. If $f \in \mathfrak{L}$ then f is finite almost everywhere.

(11.6) Remark. Propositions (5.6) and (5.8), which until now were somewhat mysterious, become quite transparent in view of (11.2) and (11.4).

(11.7) Proposition. Let f be nonnegative function on X. If $\bar{I}f = 0$, then $f \doteq 0$.

Proof. Let $\bar{I}f = 0$, $A = \{x \in X : f(x) > 0\}$, and let $g = (+\infty)f$. Since

$$\bar{I}(nf) = n\bar{I}f = 0$$

$n = 1, 2, \ldots$, and $nf \nearrow g$, we have $\bar{I}g = 0$. Because $\chi_A \leq g$, it follows that $A \in \mathfrak{I}_o$. ///

(11.8) Proposition. Let $f_n \in \mathfrak{L}$, $n = 1, 2, \ldots$, and let $\Sigma_{n=1}^{\infty} I(|f_n|) < +\infty$. Then $\Sigma_{n=1}^{\infty} |f_n| \dot{<} +\infty$. If f is a function on X such that

$$f(x) = \sum_{n=1}^{\infty} f_n(x)$$

for each $x \in X$ for which the series on the right converges, then $f \in \mathfrak{L}$ and

$$If = \sum_{n=1}^{\infty} If_n$$

<u>Proof</u>. Let $g = \Sigma_{n=1}^{\infty} |f_n|$. By (6.18),

$$Ig = \sum_{n=1}^{\infty} I(|f_n|) < +\infty$$

and we have $g \in \mathfrak{L}$. According to (11.5), $g \dot{<} +\infty$. Let

$$g_k(x) = \sum_{n=1}^{k} f_n(x)$$

for all $x \in X$ for which the right side has meaning. Then $g_k \in \mathfrak{L}$, $|g_k| \leq g$, $k = 1, 2, \ldots$, and $g_k \dot{\to} f$. Thus by (5.17) and (11.2), $f \in \mathfrak{L}$ and

$$If = \lim Ig_k = \sum_{n=1}^{\infty} If_n \quad ///$$

<u>(11.9)</u> <u>Theorem</u>. Let f be a function on X. Then $f \in \mathfrak{L}$ iff there are $f_n \in \mathfrak{F}$ and $g \in \mathfrak{L}$ such that $|f_n| \leq g$, $n = 1, 2, \ldots$, and $f_n \dot{\to} f$.

<u>Proof</u>. Let $f \in \mathfrak{L}$. Use (5.20) to choose $f_n \in \mathfrak{F}$ so that

$$I(|f - f_n|) < 2^{-n}$$

$n = 1, 2, \ldots$, and let $\varphi = \inf f_n$, $\psi = \sup f_n$, and $g = |\varphi| \vee |\psi|$. By (2.11), (2.19), their duals, and (6.5), $\varphi \in \mathfrak{L}_-$ and $\psi \in \mathfrak{L}_+$. It follows from (11.8) that

$$\sum_{n=1}^{\infty} |f - f_n| \dot{<} +\infty$$

and so $f_n \dot{\to} f$ since $|f - f_n| \dot{\to} 0$. Setting

$$A = \{x \in X : f(x) \neq \pm\infty \text{ and } f_n(x) \to f(x)\}$$

by (11.5) we have $X - A \in \mathfrak{I}_o$. For $x \in A$ either $\varphi(x) = f_n(x)$ for some integer $n \geq 1$, or $\varphi(x) = f(x)$. Thus for each $x \in A$,

$$|f(x) - \varphi(x)| \leq \sum_{n=1}^{\infty} |f(x) - f_n(x)|$$

Since $X - A \in \mathfrak{I}_o$, we obtain from (6.16), (6.15), (6.18), and (11.2) that

$$|If - I\varphi| \leq I(|f - \varphi|) \leq \sum_{n=1}^{\infty} I(|f - f_n|) < 1$$

Thus by (6.12), $\varphi \in \mathfrak{L}$. Quite analogously we can show that $\psi \in \mathfrak{L}$. Therefore $g \in \mathfrak{L}$ and, of course, $|f_n| \leq g$, $n = 1, 2, \ldots$.

The converse follows from (5.17) and (11.2). ///

Theorem (11.9) indicates that $\mathfrak{L}$ is the largest system in which $\mathfrak{F}$ is "dense." We shall formulate this precisely in exercise (11-9).

(11.10) Remark. Other than defining the null ideal $\mathfrak{I}_o$, the results of this chapter are independent of the measure space $(X, \mathfrak{I}, \iota)$. In particular, if we let

$$\mathfrak{I}_o = \{A \subset X : \bar{I}\chi_A = 0\}$$

we do not need the assumption that $\mathfrak{L}$ is a Stonian system.

Exercises

(11-1)$^+$ Let D be the Cantor discontinuum [see Ref. 16, Chapter 5, problem 0, p. 165], and let λ be Lebesgue measure [see (10-7)]. Show that

(i) $\lambda(D) = 0$.

(ii) There is a Lebesgue measurable subset of $\mathbb{R}$ which is not a Borel set. Hint: Observe that the cardinality of D is the continuum and use (7-11)(ii).

(iii) For each $\varepsilon \in (0,1)$ there is a set $D_\varepsilon \subset [0,1]$ which is homeomorphic to D and such that $\lambda(D_\varepsilon) = 1 - \varepsilon$. Hint: Construct D_ε in the same manner as D, but at the nth step remove the middle intervals of the length $\varepsilon 3^{-n}$.

(11-2) Show that there is a real-valued function f on **R** with the following properties:

(i) $f \doteq 0$ with respect to the Lebesgue measure λ [see (10-7)].

(ii) $f((a,b)) = \mathbf{R}$ for each nonempty open interval $(a,b) \subset \mathbf{R}$.

Hint: Let D be the Cantor discontinuum, and let Q be the set of all rational numbers. Observe that $\mathfrak{C} = \{x + Q : x \in D\}$ [see (10-8)(i)] is a disjoint collection of sets and that the cardinality of $\mathfrak{C}$ is the continuum. Choose a bijection $\varphi : \mathfrak{C} \to \mathbf{R}$, and let $A = \bigcup\{x + Q : x \in D\}$. For $x \in \mathbf{R}$, set $f(x) = \varphi(x + Q)$ if $x \in A$ and $f(x) = 0$ otherwise.

(11-3)*+ Let Y be an uncountable discrete space, $X = \mathbf{R} \times Y$, and let $\mathfrak{F} = C_0(X)$. Observe that if $f \in \mathfrak{F}$, then

$$\lim_{a \to +\infty} \int_{-a}^{a} f(t,y)\, dt = 0$$

for all but finitely many $y \in Y$, and set

$$If = \Sigma_{y \in Y} \lim_{a \to +\infty} \int_{-a}^{a} f(t,y)\, dt$$

for each $f \in \mathfrak{F}$. Show that I is a Bourbaki integral on $\mathfrak{F}$ and denote by $(X, \mathfrak{I}, \iota)$ the measure space induced by I. Let $\varkappa$ be the restriction of ι to the Borel σ-algebra $\mathfrak{B}$ in X. Prove

(i) If $A = \{0\} \times Y$, then $\iota(A) = +\infty$ and $\iota(F) = 0$ for each compact set $F \subset A$. Hint: Use (5-8).

(ii) $\varkappa$ is neither complete nor saturated. Hint: Use (7-12), (7-13)(iv), and the note in (3-13).

(iii) The completion $\mu = \bar{\varkappa}$ on $\mathfrak{M} = \bar{\mathfrak{B}}$ is not saturated. Hint: Let M be a subset of the Cantor discontinuum which is not a Borel set [see (11-1)(ii)]. If $A = M \times Y$, then $A \in \mathfrak{M}^\wedge - \mathfrak{M}$ [see (8-10)].

(iv) $\mathfrak{M}^\wedge = \mathfrak{I}$ and $\mu^\wedge = \iota$. Hint: Use (10.18), (10.19), and (9-7)(ii).

(v) The saturation $\varkappa^\wedge$ is not complete.

(vi) The completion of $\varkappa^\wedge$ is not saturated. Therefore, $\overline{(\varkappa^\wedge)} \neq (\bar{\varkappa})^\wedge$.

Hint: Use the hint from (iii).

(vii) For $A \in \mathfrak{I}$, let

$$\nu(A) = \sup\{\iota(F) : F \text{ compact}, F \subset A\}$$

Then ν is a complete and saturated measure on $\mathfrak{I}$ which is not regular.

(viii) $\nu \leq \iota$ and $\nu(A) = \iota(A)$ for each $A \in \mathfrak{I}$ which is either open or ι-σ-finite.

(11-4) Denote by Q the set of all rational numbers from $(0,1]$. Let $X = [0,1]$ and let $\mathfrak{F}$ consist of all functions f on X satisfying the following conditions:

(a) On Q, f is the restriction of a linear function vanishing at zero

(b) On X - Q, f is the restriction of a function from $\mathcal{C}(X)$.

Show that $\mathfrak{F}$ is a fundamental system on X which is not Stonian [see (10.1)].

If $f \in \mathfrak{F}$, find a $g \in \mathcal{C}(X)$ so that $f(x) = g(x)$ for all $x \in X - Q$, and set

$$I_1 f = \int_0^1 g(x)\, dx$$

Furthermore, let I_2 and I_3 be the Dirac integrals [see (4-1)] on $\mathfrak{F}$ at 1 and $\sqrt{2}/2$, respectively. Prove

(i) I_1 is a well-defined Bourbaki integral on $\mathfrak{F}$.

(ii) Every bounded Baire function on X [see (3-13)] belongs to $\mathfrak{L}_1$.

Hint: Show that $\mathcal{C}(X) \subset \mathfrak{L}_1$.

(iii) $\mathfrak{L}_1$ is a Stonian system.

(iv) If $(X, \mathfrak{I}_1, \iota_1)$ is the measure space induced by I_1, then

$\mathfrak{I}_1 = \{A \in \Lambda : A \subset [0,1]\}$ and $\iota_1(A) = \lambda(A)$ for each $A \in \mathfrak{I}_1$ [see (10-7)].

In particular, $Q \in \mathfrak{I}_1$ and $\iota_1(Q) = 0$.

(v) $f \in \mathfrak{L}_2$ iff f restricted to Q is a linear function vanishing at zero. In particular, $\mathfrak{L}_2$ is not a Stonian system.

(vi) $\mathfrak{L}_3$ consist of all functions f on X for which $f(\sqrt{2}/2) \neq \pm\infty$. In particular, $\mathfrak{L}_3$ is a Stonian system.

In exercises (11-5) through (11-11) we shall assume the same general situation which was set up at the beginning of this chapter. By $\mathcal{J}$ we shall denote the family of all measurable functions on X, and by $\mathcal{L}_o$ we shall denote the family of all finite summable functions on X.

(11-5)* Let $\{f_n\} \subset \mathcal{J}$, $\{g_n\} \subset \mathcal{L}$, $g \in \mathcal{L}$, and let $|f_n| \dot{\leq} g_n$, n = 1, 2, ... Furthermore, let $g_n \dot{\to} g$, $Ig_n \to Ig$, and let $f_n \dot{\to} f$ for some function f on X. Prove that $f \in \mathcal{L}$ and $If_n \to If$. Hint: Observe that $f_n \in \mathcal{L}$ and that you can assume $\{g_n\} \subset \mathcal{L}_o$ and $g \in \mathcal{L}_o$. Using (11.2), apply (5.10) to the sequences $\{g_n - f_n\}$ and $\{g_n + f_n\}$.

(11-6) Let $\{f_n\} \subset \mathcal{L}_o$, $f \in \mathcal{L}_o$, and let $f_n \dot{\to} f$. Prove that

$$I(|f_n - f|) \to 0 \text{ iff } I(|f_n|) \to I(|f|)$$

Hint: Use (11-5).

(11-7)* Let $\{f_n\} \subset \mathcal{L}_o$ and $\Sigma_{n=1}^{\infty} I(|f_{n+1} - f_n|) < +\infty$. Prove that $f_n \dot{\to} f$ for some $f \in \mathcal{L}_o$ and that $I(|f_n - f|) \to 0$. Hint: Use (11.8) to show that there is an $f \in \mathcal{L}_o$ such that

$$f \doteq f_1 + \sum_{n=1}^{\infty} (f_{n+1} - f_n) = \lim f_n$$

Observe that

$$\begin{aligned}\lim_{n\to+\infty} I(|f - f_n|) &= \lim_{n\to+\infty} I(\lim_{k\to+\infty} |f_k - f_n|) \\ &\leq \lim_{n\to+\infty} \lim_{k\to+\infty} I(|f_k - f_n|) \\ &\leq \lim_{n\to+\infty} \sum_{i=n}^{\infty} I(|f_{i+1} - f_i|) = 0\end{aligned}$$

(11-8) Let $\{f_n\} \subset \mathcal{J}$, and let $I(|f_n|) \to 0$. Prove

(i) There is a subsequence $\{g_n\}$ of $\{f_n\}$ such that $g_n \dot{\to} 0$.

Hint: Choose $\{g_n\}$ so that $I(|g_n|) < 2^{-n}$ and use (11.8).

(ii) Show by example that $\{f_n(x)\}$ may converge to zero at no $x \in X$.

Hint: Let $X = [0,1]$, $\mathfrak{F} = \mathfrak{C}(X)$, and let $If = \int_0^1 f(x)\,dx$ for each $f \in \mathfrak{F}$.

For $n = 0, 1, \ldots$ and $k = 1, 2, \ldots, 2^n$ let $I_{n,k} = [(k-1)2^{-n}, k2^{-n}]$ and $f_{n,k} = \chi_{I_{n,k}}$. Consider the sequence

$$f_{0,1}, f_{1,1}, f_{1,2}, f_{2,1}, f_{2,2}, f_{2,3}, f_{2,4}, f_{3,1}, \ldots$$

(Draw a picture.)

(11-9)$^+$ (i) For $f,g \in \mathfrak{L}_o$, set

$$\tau(f,g) = I(|f - g|)$$

and show that τ is a pseudometric on $\mathfrak{L}_o$ (see Ref. 16, Chapter 4, p. 119).

(ii) For $f,g \in \mathfrak{L}_o$, write $f \sim g$ iff $\tau(f,g) = 0$. Show that $\sim$ is an equivalence relation on $\mathfrak{L}_o$ and denote by $[f]$ the equivalence class determined by $f \in \mathfrak{L}_o$. If $\mathfrak{G} \subset \mathfrak{L}_o$, set $[\mathfrak{G}] = \{[f] : f \in \mathfrak{G}\}$.

(iii) For $f,g \in \mathfrak{L}_o$, show that $f \sim g$ iff $f \doteq g$, and let

$$\rho([f],[g]) = \tau(f,g)$$

Show that ρ is a well-defined metric on $[\mathfrak{L}_o]$.

(iv) Prove that in the topology given by ρ, $[\mathfrak{F}]$ is a dense subset of $[\mathfrak{L}_o]$. Hint: Use (5.20).

(v) The set $[\mathfrak{L}_o]$ with the metric ρ is a complete metric space.

Hint: If $\{[f_n]\} \subset [\mathfrak{L}_o]$ is a Cauchy sequence, find a subsequence $\{[g_n]\}$ so that $I(|g_{n+1} - g_n|) < 2^{-n}$ and use (11-7). Observe, that a Cauchy sequence converges iff it has a convergent subsequence

(vi) Conclude from (iv) and (v) that the metric space $([\mathfrak{L}_o],\rho)$ is the completion of the metric space $([\mathfrak{F}],\rho)$ [see Ref. 13, (6.85), p. 77].

(11-10)$^+$ Let $f \in \mathfrak{J}$, and let $A = \{x \in X : f(x) \neq 0\}$. Prove that A is ι-σ-finite iff there is a sequence $\{f_n\} \subset \mathfrak{F}$ such that $f_n \dot{\to} f$. Hint: If A is ι-σ-finite, then there is $\{A_n\} \subset \mathfrak{I}$ such that $A_n \subset A_{n+1}$,

$\iota(A_n) < +\infty$, $n = 1, 2, \ldots$, and $\bigcup_{n=1}^{\infty} A_n = A$. Letting $g_n = (f\chi_{A_n} \wedge n) \vee (-n)$, we have $g_n \in \mathfrak{L}$, $n = 1, 2, \ldots$, and $g_n \to f$. By (5.20) there are $f_n \in \mathfrak{F}$ for which $I(|f_n - g_n|) < 2^{-n}$. It follows from (11.8) that $f_n - g_n \dot{\to} 0$. To show the converse, it suffices to use (10.8) and observe that, up to an ι-null set, the set A is contained in $\bigcup_{n=1}^{\infty} \{x \in X : f_n(x) \neq 0\}$.

<u>(11-11)</u>*+ <u>Luzin's theorem</u>. Suppose that X is a locally compact Hausdorff space, $\mathfrak{F} = C_0(X)$, and I is a Bourbaki integral on $\mathfrak{F}$. Let $f \in \mathfrak{J}$ be finite, and let $\iota(\{x \in X : f(x) \neq 0\}) < +\infty$. Then for every $\varepsilon > 0$ there is a $g \in \mathfrak{F}$ such that $\iota(\{x \in X : f(x) \neq g(x)\}) < \varepsilon$.

(i) Prove Luzin's theorem. <u>Hint</u>: Let $A = \{x \in X : f(x) \neq 0\}$. Given $\varepsilon > 0$, choose a compact set F and an open set G with G^- compact so that $F \subset A$, $F \subset G$, $\iota(A - F) < \varepsilon/3$, and $\iota(G - F) < \varepsilon/3$ [see (10.19)]. By (11-10) there is $\{f_n\} \subset \mathfrak{F}$ for which $f_n \dot{\to} f$. Using (8-8) and (10.19), we can find an open set H such that $\iota(H) < \varepsilon/3$ and $f_n \to f$ uniformly on F - H. Thus f is continuous on F - H. Combining the Tietze extension theorem (see Ref. 16, Chapter 7, problem O, p. 242) and Urysohn's lemma construct a function $g \in C(X)$ so that supp $g \subset G$ and $g(x) = f(x)$ for each $x \in F - H$.

(ii) Use exercise (10-2) with an infinite X to show that the assumption $\iota(\{x \in X : f(x) \neq 0\}) < +\infty$ is essential in Luzin's theorem.

<u>(11-12)</u>* Let X be a compact Hausdorff space. Using the technique from (10-12), show that

(i) If X has no isolated points, then $\|X\| \geq \mathfrak{c}$.

(ii) If X is uncountable and metrizable, then it contains a homeomorphic image of the Cantor discontinuum [see Ref. 16, Chapter 5, problem O, p. 165].

<u>(11-13)</u>*+ Let X be a locally compact metrizable space. Recall the definition of a diffused measure [see (8-11)], and prove the following:

(i) If $\|X\| > \aleph_0$ and X contains a condensation point [see (10-12)(i)], then there is a diffused, regular, Borel measure μ in X such that $\mu(X) = 1$. <u>Hint</u>: Use (11-12)(ii) and (11-1)(iii).

(ii) If X is separable and contains no isolated points, then there is a diffused, regular, Borel measure μ in X such that $\mu(X) = 1$ and supp $\mu = X$ [see (9-8)]. <u>Hint</u>: Choose a countable open base $\{G_n\}$ for the topology of X. Using (11-12)(i), apply (i) to each G_n.

(iii) If X contains no isolated points, then there is a diffused, regular, Borel measure μ in X such that $\mu(X) > 0$ and supp $\mu = X$. <u>Hint</u>: Use (ii) and the fact that X is a free union of separable spaces [see Ref. 8, Chapter IX, Theorem 5.3, p. 186 and Chapter XI, Theorem 7.3, p. 241].

<u>(11-14)</u>* Let A be the set of all real numbers whose decimal expansions contain at least one digit only finitely many times. Show that $A \in \Lambda$ and $\lambda(A) = 0$ [see (10-7)]. <u>Hint</u>: Choose a digit i and let A_i be the set of all real numbers from the interval [0, 1] whose decimal expansions contain no digit i. Show that $A_i \in \Lambda$ and $\lambda(A_i) = 0$.

<u>Note</u>: An interesting generalization of this exercise was given by F. Rubin [see The measure of recognizable sets of real numbers, <u>Amer. Math. Monthly</u>, 83(1976), 348-349].

12. INTEGRALS INDUCED BY MEASURES

Given a measure space $(X,\mathfrak{M},\mu)$, we shall define a Stonian fundamental system $\mathfrak{F}$ on X and a Daniell integral I on $\mathfrak{F}$ which we extend to $\bar{\mathfrak{L}}$. We show that the measure ι induced by I is an extension of μ and that I coincides with the abstract Lebesgue integral with respect to ι [see (8.17)].

Throughout this chapter we shall assume that $(X,\mathfrak{M},\mu)$ is an arbitrary measure space. By $\mathfrak{F}$ we shall denote the family of all simple functions f on X [see (7.25)] such that

$$\mu(\{x \in X : f(x) \neq 0\}) < +\infty$$

From (7.13), (7.14), (7.16), and (7.10) we immediately obtain the following:

(12.1) $\mathfrak{F}$ is a fundamental system on X.

(12.2) If $f \in \mathfrak{F}$ and if $g \geq 0$ is a simple function, then $f \wedge g \in \mathfrak{F}$. In particular, the system $\mathfrak{F}$ is Stonian.

Let $f \in \mathfrak{F}$, and let

$$f = \sum_{i=1}^{n} a_i \chi_{A_i}$$

be its canonical form [see (7.27)]. Setting

$$If = \sum_{i=1}^{n} a_i \mu(A_i)$$

we obtain a well-defined map

$$I : \mathfrak{F} \to \mathbf{R}$$

[see (7.28)] with the following properties:

(12.3) $I(af) = aIf$ for each $f \in \mathfrak{F}$ and each $a \in \mathbf{R}$.

(12.4) $(f \in \mathfrak{F},\ f \geq 0) \Rightarrow If \geq 0$.

In order to show that I is also additive on $\mathfrak{F}$, i.e.,

$$I(f + g) = If + Ig$$

for all $f, g \in \mathfrak{F}$, we shall need some preliminary results.

(12.5) Definition. Let $\mathfrak{C}$ and $\mathfrak{D}$ be families of sets such that

$$\bigcup\{C : C \in \mathfrak{C}\} = \bigcup\{D : D \in \mathfrak{D}\}$$

We say that $\mathfrak{D}$ is a refinement of $\mathfrak{C}$ whenever for each $D \in \mathfrak{D}$ there is a $C \in \mathfrak{C}$ containing D. We say that $\mathfrak{D}$ is a strong refinement of $\mathfrak{C}$ whenever for each $C \in \mathfrak{C}$ and each $D \in \mathfrak{D}$ either $C \cap D = \emptyset$ or $D \subset C$.

If $\mathfrak{C}$ is a collection of sets, the following facts are obvious:

(i) Any strong refinement of $\mathfrak{C}$ is a refinement of $\mathfrak{C}$.
(ii) $\mathfrak{C}$ is a refinement of itself.
(iii) $\mathfrak{C}$ is a strong refinement of itself iff it is disjoint.

(12.6) Lemma. Every family $\{A_i\}_{i=1}^n \subset \mathfrak{M}$ has a disjoint strong refinement $\{B_j\}_{j=1}^m \subset \mathfrak{M}$.

Proof. Let $\{A_i\}_{i=1}^n \subset \mathfrak{M}$. Since the lemma holds trivially for $n = 1$, suppose that it holds for an integer $n - 1 \geq 1$. Then there is a disjoint strong refinement $\{C_j\}_{j=1}^p \subset \mathfrak{M}$ of $\{A_i\}_{i=1}^{n-1}$. Letting $B_j = C_j \cap A_n$, $B_{p+j} = C_j - A_n$, $j = 1, \ldots, p$, and $B_{2p+1} = A_n - \bigcup_{i=1}^{n-1} A_i$, we see that $\{B_j\}_{j=1}^{2p+1} \subset \mathfrak{M}$ and that it is a disjoint strong refinement of $\{A_i\}_{i=1}^n$. ///

(12.7) Lemma. Let $f = \Sigma_{i=1}^n a_i \chi_{A_i}$, where $a_i \in \mathbf{R}$, $A_i \in \mathfrak{M}$, and $\mu(A_i) < +\infty$ whenever $a_i \neq 0$, $i = 1, \ldots, n$. Then $f \in \mathfrak{F}$ and

$$If = \sum_{i=1}^{n} a_i \mu(A_i)$$

Proof. Because $\mu(A_i) < +\infty$ whenever $a_i \neq 0$, f belongs to $\mathfrak{F}$ and the sum $\Sigma_{i=1}^n a_i \mu(A_i)$ has meaning. Let

$$f = \sum_{j=1}^{m} b_j \chi_{B_j}$$

be the canonical form of f, and let $\{C_k\}_{k=1}^{p} \subset \mathfrak{M}$ be a disjoint strong refinement of $\{A_1, \ldots, A_n, B_1, \ldots, B_m\}$. We may assume that $C_k \neq \emptyset$, k = 1, ..., p. There are numbers $a_{i,k}$ and $b_{j,k}$ equal to 0 or 1, i = 1, ..., n, j = 1, ..., m, k = 1, ..., p, such that

$$\chi_{A_i} = \sum_{k=1}^{p} a_{i,k}\chi_{C_k} \quad \text{and} \quad \chi_{B_j} = \sum_{k=1}^{p} b_{j,k}\chi_{C_k}$$

Hence

$$f = \sum_{i=1}^{n} a_i\chi_{A_i} = \sum_{k=1}^{p} \left(\sum_{i=1}^{n} a_i a_{i,k}\right)\chi_{C_k}$$

and also

$$f = \sum_{j=1}^{m} b_j\chi_{B_j} = \sum_{k=1}^{p} \left(\sum_{j=1}^{m} b_j b_{j,k}\right)\chi_{C_k}$$

Since $C_1, \ldots, C_p$ are disjoint and nonempty, by evaluating f at an $x \in C_k$ we obtain

$$\sum_{i=1}^{n} a_i a_{i,k} = \sum_{j=1}^{m} b_j b_{j,k}$$

k = 1, ..., p. Therefore,

$$\begin{aligned} If &= \sum_{j=1}^{m} b_j\mu(B_j) = \sum_{j=1}^{m} b_j\left(\sum_{k=1}^{p} b_{j,k}\mu(C_k)\right) \\ &= \sum_{k=1}^{p} \left(\sum_{j=1}^{m} b_j b_{j,k}\right)\mu(C_k) = \sum_{k=1}^{p} \left(\sum_{i=1}^{n} a_i a_{i,k}\right)\mu(C_k) \\ &= \sum_{i=1}^{n} a_i\left(\sum_{k=1}^{p} a_{i,k}\mu(C_k)\right) = \sum_{i=1}^{n} a_i\mu(A_i) \quad /// \end{aligned}$$

(12.8) Corollary. If $f, g \in \mathfrak{F}$, then

$$I(f + g) = If + Ig$$

(12.9) Proposition. I is a Daniell integral on $\mathfrak{F}$.

Proof. According to (12.3), (12.4), and (12.8), I is a nonnegative linear functional on $\mathfrak{F}$. Thus it suffices to show that $If_n \to 0$ whenever

$\{f_n\} \subset \mathfrak{F}$ and $f_n \searrow 0$. Choose $\{f_n\} \subset \mathfrak{F}$ so that $f_n \searrow 0$ and choose $\epsilon > 0$. Letting

$$\alpha = \sup\{f_1(x) : x \in X\}$$

$$A = \{x \in X : f_1(x) \neq 0\}$$

$$A_n = \{x \in A : f_n(x) < \epsilon\}$$

$n = 1, 2, \ldots$, we have $\alpha \in \mathbf{R}$, A and A_n belong to $\mathfrak{M}$, $A_n \subset A_{n+1}$, $A = \bigcup_{n=1}^{\infty} A_n$, and $\mu(A) < +\infty$. Since for $n = 1, 2, \ldots$,

$$\begin{aligned} If_n &= I(f_n\chi_{A_n}) + I(f_n\chi_{A-A_n}) \leq \epsilon I\chi_{A_n} + \alpha I\chi_{A-A_n} \\ &= \epsilon\mu(A_n) + \alpha[\mu(A) - \mu(A_n)] \end{aligned}$$

from (8.5) we obtain that

$$\lim If_n \leq \epsilon\mu(A)$$

It follows from the arbitrariness of ϵ that $If_n \to 0$. ///

Using the methods of Chapters 2, 4-6, and 10 we can extend the integral I to $\bar{\mathfrak{L}}$ and define a new measure space $(X, \mathfrak{J}, \iota)$ induced by I.

(12.10) Definition. The integral I and the measure space $(X, \mathfrak{J}, \iota)$ induced by I are said to be induced by the measure μ.

The families of all $\mathfrak{M}$-measurable and $\mathfrak{J}$-measurable functions on X are denoted by $\mathfrak{m}$ and $\mathfrak{I}$, respectively.

We shall proceed to study the relationship between the measures μ and ι.

(12.11) Lemma. Let $g \in \mathfrak{F}_+$, $g \geq 0$, and let $A = \{x \in X : g(x) \geq 1\}$. Then $A \in \mathfrak{M}$ and $\mu(A) \leq Ig$.

Proof. There is a sequence $\{g_n\} \subset \mathfrak{F}$ such that $g_n \nearrow g$. Thus $g \in \mathfrak{m}$, and consequently, $A \in \mathfrak{M}$. Let

$$A_n = \{x \in A : g_n(x) > 0\} \qquad (n = 1, 2, \ldots)$$

Because $A_n \in \mathfrak{M}$ and $\mu(A_n) < +\infty$, we have $\chi_{A_n} \in \mathfrak{F}$. Clearly, $\chi_{A_n} \leq g$, and so

$$\mu(A_n) = I\chi_{A_n} \leq Ig \qquad (n = 1, 2, \ldots)$$

Passing to the limit, we obtain that $\mu(A) \leq Ig$; for $A_n \subset A_{n+1}$ and $A = \bigcup_{n=1}^{\infty} A_n$. ///

(12.12) Lemma. Let $A \in \mathfrak{I}$ and $\iota(A) < +\infty$. Then there are $B, C \in \mathfrak{M}$ such that $B \subset A \subset C$, $\mu(C) < +\infty$, and $\mu(B - C) = 0$.

Proof. (a) Given $\epsilon > 0$ there are $f \in \mathfrak{F}_-$ and $g \in \mathfrak{F}_+$ such that

$$0 \leq f \leq \chi_A \leq g$$

$Ig < +\infty$, and $I(g - f) < \epsilon$. Because $f, g \in \mathfrak{m}$, the sets

$$B_\epsilon = \{x \in X : f(x) > 0\} \qquad \text{and} \qquad C_\epsilon = \{x \in X : g(x) \geq 1\}$$

belong to $\mathfrak{M}$. Clearly, $B_\epsilon \subset A \subset C_\epsilon$. Since

$$C_\epsilon - B_\epsilon = \{x \in X : g(x) - f(x) \geq 1\}$$

and since g and $g - f$ belong to $\mathfrak{F}_+$, it follows from (12.11) that

$$\mu(C_\epsilon) \leq Ig < +\infty \qquad \text{and} \qquad \mu(C_\epsilon - B_\epsilon) \leq I(f - g) < \epsilon$$

(b) Let $\epsilon_n = 1/n$, $n = 1, 2, \ldots$, $B = \bigcup_{n=1}^{\infty} B_{\epsilon_n}$, and $C = \bigcap_{n=1}^{\infty} C_{\epsilon_n}$. Then $B, C \in \mathfrak{M}$, $B \subset A \subset C$, and $\mu(C) \leq \mu(C_{\epsilon_1}) < +\infty$. Since

$$\mu(C - B) \leq \mu(C_{\epsilon_n} - B_{\epsilon_n}) < \frac{1}{n}$$

$n = 1, 2, \ldots$, it follows that $\mu(C - B) = 0$. ///

(12.13) Corollary. Let $A \in \mathfrak{I}$ be ι-σ-finite. Then there are $B, C \in \mathfrak{M}$ such that $B \subset A \subset C$, C is μ-σ-finite and $\mu(C - B) = 0$.

Proof. There are $A_n \in \mathfrak{I}$ such that $\iota(A_n) < +\infty$, $n = 1, 2, \ldots$, and $A = \bigcup_{n=1}^{\infty} A_n$. By (12.12) there are $B_n, C_n \in \mathfrak{M}$ such that $B_n \subset A_n \subset C_n$, $\mu(C_n) < +\infty$, and $\mu(C_n - B_n) = 0$, $n = 1, 2, \ldots$. It suffices to let $B = \bigcup_{n=1}^{\infty} B_n$ and $C = \bigcup_{n=1}^{\infty} C_n$. ///

(12.14) Theorem. $\mathfrak{M} \subset \mathfrak{I}$ and $\iota(A) = \mu(A)$ for each $A \in \mathfrak{M}$.

Proof. (a) From (12.2) we obtain that $A \in \mathfrak{M}$, and $f \in \mathfrak{F}$ implies $\chi_A \wedge f \in \mathfrak{F}$. Therefore, $\mathfrak{M} \subset \mathfrak{I}$.

(b) Let $A \in \mathfrak{M}$. If $\mu(A) < +\infty$, then $\chi_A \in \mathfrak{F}$ and $\mu(A) = I\chi_A = \iota(A)$. If $\mu(A) = +\infty$, then by (12.12) also $\iota(A) = +\infty$. ///

Because the measure ι is complete and saturated [see (10.18)] we obtain the following corollary.

(12.15) Corollary. Any measure can be extended to a complete saturated measure.

We note, however, that a complete and saturated extension of a measure is, in general, not unique. A careful analysis of this problem is given in exercises (8-9), (8-10), and (9-6).

(12.16) Proposition. Let μ be a complete measure. If $A \in \mathfrak{I}$ is ι-σ-finite, then $A \in \mathfrak{M}$.

Proof. Let $A \in \mathfrak{I}$ be ι-σ-finite. By (12.13) there are $B, C \in \mathfrak{M}$ such that $B \subset A \subset C$ and $\mu(C - B) = 0$. By the completeness of μ, $A = B \cup (A - B)$ belongs to $\mathfrak{M}$. ///

(12.17) Theorem. Let μ be a complete and saturated measure. Then $(X, \mathfrak{M}, \mu) = (X, \mathfrak{I}, \iota)$.

Proof. According to (12.14), it suffices to show that $\mathfrak{I} \subset \mathfrak{M}$. Let $A \in \mathfrak{I}$, and let $B \in \mathfrak{M}$, $\mu(B) < +\infty$. By (12.14), $A \cap B \in \mathfrak{I}$ and

$$\iota(A \cap B) \leq \iota(B) = \mu(B) < +\infty$$

Since μ is complete, $A \cap B \in \mathfrak{M}$ [see (12.16)]. Thus $A \in \mathfrak{M}$, for μ is saturated. ///

(12.18) Theorem. Let I_1 be an integral (Bourbaki or Daniell) on a fundamental system $\mathfrak{F}_1$ on X, and suppose that $\mathfrak{L}_1$ is a Stonian system. Let $(X, \mathfrak{I}_1, \iota_1)$ be the measure space induced by I_1. If I_2 is the integral induced by ι_1, then $\bar{\mathfrak{L}}_1 = \bar{\mathfrak{L}}_2$ and $I_1 f = I_2 f$ for each $f \in \bar{\mathfrak{L}}_1$.

Proof. (a) By (10.18) the measure ι_1 is complete and saturated and so by (12.17), $(X, \mathfrak{I}_1, \iota_1)$ is the measure space induced by both I_1 and I_2. It follows that if $f \geq 0$ is a $\mathfrak{I}_1$-simple function on X, then

$f \in \mathfrak{L}_{1+} \cap \mathfrak{L}_{2+}$ and $I_1 f = I_2 f$.

(b) Let $f \geq 0$ be a $\mathfrak{I}_1$-measurable function on X. By (7.29) there is a sequence $\{f_n\}$ of nonnegative $\mathfrak{I}_1$-simple functions on X such that $f_n \nearrow f$. Thus by (a), $f \in \mathfrak{L}_{1+} \cap \mathfrak{L}_{2+}$ and

$$I_1 f = \lim I_1 f_n = \lim I_2 f_n = I_2 f$$

(c) Let f be an arbitrary $\mathfrak{I}_1$-measurable function on X. According to (b), $f^+, f^- \in \mathfrak{L}_{1+} \cap \mathfrak{L}_{2+}$ and

$$I_1 f^+ = I_2 f^+ \qquad I_1 f^- = I_2 f^-$$

It follows from (10.15) that $f \in \bar{\mathfrak{L}}_1$ iff $f \in \bar{\mathfrak{L}}_2$. If $f \in \bar{\mathfrak{L}}_1$, then

$$I_1 f = I_1 f^+ - I_1 f^- = I_2 f^+ - I_2 f^- = I_2 f \quad ///$$

At this point we have come to a full circle in our process of inducing measures by integrals and vice versa. The following summarizes what the process does.

(i) A measure μ induces an integral I on a Stonian fundamental system. This integral, in turn, induces a measure ι which is a complete and saturated extension of μ. Moreover, if it so happens that the measure μ is already complete and saturated, then $\iota = \mu$.

(ii) An integral I for which the system of I-summable functions is Stonian induces a measure ι. This measure induces back the integral I.

We shall close this chapter by establishing the relationship between the integral I induced by the measure μ and the abstract Lebesgue integral with respect to μ [see (8.17)].

By $\bar{\mathfrak{L}}_\mu$ we shall denote the family of all $f \in \mathfrak{M}$ for which $\int f \, d\mu$ exists, and we let

$$\mathfrak{L}_\mu = \{f \in \bar{\mathfrak{L}}_\mu : \int f \, d\mu \neq \pm\infty\}$$

(12.19) Theorem. $\bar{\mathfrak{L}}_\mu = \bar{\mathfrak{L}} \cap \mathfrak{m}$ and $\int f\,d\mu = If$ for each $f \in \bar{\mathfrak{L}}_\mu$.

Proof. Denote by $\mathfrak{S}_+$ the family of all nonnegative $\mathfrak{M}$-simple functions. Clearly, $\mathfrak{S}_+ \subset \bar{\mathfrak{L}} \cap \bar{\mathfrak{L}}_\mu$.

(a) Let $f \in \mathfrak{S}_+$, and let $f = \Sigma_{i=1}^n a_i \chi_{A_i}$ be the canonical form of f. From (12.14) we obtain

$$If = \sum_{i=1}^{n} a_i I\chi_{A_i} = \sum_{i=1}^{n} a_i \iota(A_i) = \sum_{i=1}^{n} a_i \mu(A_i) = \int f\,d\mu$$

(b) Let $f \in \mathfrak{m}$ and $f \geq 0$. Then there is a sequence $\{g_n\} \subset \mathfrak{S}_+$ such that $g_n \nearrow f$. Thus by (6.17) and (a),

$$If = \sup Ig_n = \sup\{Ig : g \in \mathfrak{S}_+,\ g \leq f\}$$

$$= \sup\{\int g\,d\mu : g \in \mathfrak{S}_+,\ g \leq f\} = \int f\,d\mu$$

(c) Let $f \in \mathfrak{m}$ be arbitrary. According to (b),

$$If^+ = \int f^+\,d\mu \quad \text{and} \quad If^- = \int f^-\,d\mu$$

Thus $If^+ - If^-$ has meaning iff $\int f^+\,d\mu - \int f^-\,d\mu$ does. It follows from (10.15) that $f \in \bar{\mathfrak{L}}$ iff $f \in \bar{\mathfrak{L}}_\mu$. If $f \in \bar{\mathfrak{L}}_\mu$, then

$$If = If^+ - If^- = \int f^+\,d\mu - \int f^-\,d\mu = \int f\,d\mu \quad ///$$

The previous theorem is a very important tool by which we can readily obtain all basic properties of the abstract Lebesgue integral. We shall tacitly ask the reader to apply it at many places in the sequel.

(12.20) Corollary. Let μ be a complete measure. Then $\mathfrak{L}_\mu = \mathfrak{L}$.

Proof. By (12.19), $\mathfrak{L}_\mu = \mathfrak{L} \cap \mathfrak{m}$ and so it suffices to show that $\mathfrak{L} \subset \mathfrak{m}$.

(a) Let $f \in \mathfrak{L}$, $f \geq 0$, and for $c \in \mathbf{R}$, let $A_c = \{x \in X : f(x) > c\}$. By (10.7), $A_c \in \mathfrak{J}$ and $\iota(A_c) < +\infty$ for each $c > 0$. Thus by (12.16), $A_c \in \mathfrak{M}$ for each $c > 0$. Since $A_c = X$ whenever $c < 0$ and $A_0 = \bigcup_{n=1}^{\infty} A_{1/n}$, we have $f \in \mathfrak{m}$.

(b) Because $f \in \mathfrak{L}$ iff $f^+, f^- \in \mathfrak{L}$ and $f \in \mathfrak{m}$ iff $f^+, f^- \in \mathfrak{m}$, it follows from (a) that $\mathfrak{L} \subset \mathfrak{m}$. ///

(12.21) Corollary. Let μ be a complete and saturated measure. Then $\bar{\mathfrak{L}}_\mu = \bar{\mathfrak{L}}$.

Proof. By (12.17) and (10.13), $\bar{\mathfrak{L}} \subset \mathfrak{m}$. The corollary follows from (12.19). ///

(12.22) Corollary. Let J be an integral (Bourbaki or Daniell) on a fundamental system $\mathcal{E}$ on X and let the system of J-summable functions be Stonian. If $(X, \mathfrak{N}, \nu)$ is the measure space induced by J, then $\bar{\mathfrak{L}}_\nu$ consists of all J-integrable functions and $\int f\, d\nu = Jf$ for each $f \in \bar{\mathfrak{L}}_\nu$.

This corollary follows directly from (12.18), (12.19), (10.18), and (12.21).

Exercises

(12-1) Let $X = \mathbf{R}$ and let Λ be the σ-algebra of all Lebesgue measurable subsets of X [see (10-7)]. Show that

(i) The integral induced by the Lebesgue measure λ on Λ is not a Bourbaki integral.

(ii) The integral induced by the counting measure on Λ [see (8-1)(iv)] is a Bourbaki integral.

(iii) The integral induced by the Dirac measure at $x_o \in X$ on Λ [see (8-1)(v)] is a Bourbaki integral.

(12-2) Let $(X, \mathfrak{M}, \mu)$ be the measure space defined in (9-10), and let I be the integral induced by the measure μ. Show that

(i) I is not a Bourbaki integral.

(ii) $(X, \mathfrak{M}, \mu)$ is the measure space induced by I.

The remaining exercises will be concerned with an important construction of measures due to Carathéodory.

Let X be a set. A nonnegative function μ_e on exp X is called an outer (or exterior) measure on X whenever

(a) $\mu_e(\emptyset) = 0$

(b) $A \subset B \subset X \Rightarrow \mu_e(A) \leq \mu_e(B)$

(c) If $\{A_n\}$ is a countable family of subsets of X, then

$$\mu_e(\bigcup_n A_n) \leq \sum_n \mu_e(A_n)$$

A set $A \subset X$ is called μ_e-measurable if

$$\mu_e(M) = \mu_e(M \cap A) + \mu_e(M - A)$$

for each set $M \subset X$.

(12-3) Let X be a set. For $A \subset X$, set $\mu_e(A) = 0$ if A is countable and $\mu_e(A) = 1$ otherwise. Show that

(i) μ_e is an outer measure on X.

(ii) μ_e is a measure on exp X iff X is countable.

(iii) A set $A \subset X$ is μ_e-measurable iff A or X - A is countable.

(iv) Let $\mathfrak{M}_e$ be the family of all μ_e-measurable subsets of X. Then $(X, \mathfrak{M}_e, \mu_e)$ is a measure space.

In exercises (12-4) through (12-7) we shall assume that μ_e is an outer measure on a set X, and we shall denote by $\mathfrak{M}_e$ the family of all μ_e-measurable subsets of X.

(12-4) Show that $A \in \mathfrak{M}_e$ iff

$$\mu_e(M) \geq \mu_e(M \cap A) + \mu_e(M - A)$$

for each $M \subset X$.

(12-5)* Show that if $A, B \in \mathfrak{M}_e$, then also $A - B \in \mathfrak{M}_e$. Hint: Choose $M \subset X$, and let $P = M \cap (A - B)$ and $Q = M - (A - B)$. Observe that

$$\mu_e(Q) = \mu_e(Q \cap B) + \mu_e(Q - B) = \mu_e(M \cap B) + \mu_e(Q - B)$$

$$\mu_e(M - B) = \mu_e[(M - B) \cap A] + \mu_e[(M - B) - A] = \mu_e(P) + \mu_e(Q - B)$$

$$\mu_e(M) = \mu_e(M \cap B) + \mu_e(M - B)$$

(12-6)* Let $\{A_n\}_{n=1}^{\infty} \subset \mathfrak{M}_e$ be a disjoint countable family, $B_n = \bigcup_{i=1}^{n} A_i$, and let $A = \bigcup_{n=1}^{\infty} A_n$. Prove

(i) $B_n \in \mathfrak{M}_e$ and

$$\mu_e(M) = \sum_{i=1}^{n} \mu_e(M \cap A_i) + \mu_e(M - B_n)$$

for each $M \subset X$, $n = 1, 2, \ldots$. Hint: Assume true for $n = k$ and observe that

$$\begin{aligned}
\mu_e(M) &= \mu_e(M \cap A_{k+1}) + \mu_e(M - A_{k+1}) \\
&= \mu_e(M \cap A_{k+1}) + \{\mu_e[(M - A_{k+1}) \cap B_k] + \mu_e[(M - A_{k+1}) - B_k]\} \\
&= \mu_e(M \cap A_{k+1}) + \mu_e(M \cap B_k) + \mu_e(M - B_{k+1}) \\
&= \mu_e(M \cap A_{k+1}) + \{\sum_{i=1}^{k} \mu_e[(M \cap B_k) \cap A_i] \\
&\quad + \mu_e[(M \cap B_k) - B_k]\} + \mu_e(M - B_{k+1}) \\
&= \sum_{i=1}^{k+1} \mu_e(M \cap A_i) + \mu_e(M - B_{k+1}) \geq \mu_e(M \cap B_{k+1}) \\
&\quad + \mu_e(M - B_{k+1}) \geq \mu_e(M)
\end{aligned}$$

(ii) $A \in \mathfrak{M}_e$ and

$$\mu_e(M) = \sum_{n=1}^{\infty} \mu_e(M \cap A_n) + \mu_e(M - A)$$

for each $M \subset X$. Hint: Using the inequality

$$\mu_e(M - B_n) \geq \mu_e(M - A)$$

$n = 1, 2, \ldots,$ and passing to the limit in (i) we obtain

$$\begin{aligned}
\mu_e(M) &\geq \sum_{n=1}^{\infty} \mu_e(M \cap A_n) + \mu_e(M - A) \\
&\geq \mu_e(M \cap A) + \mu_e(M - A) \geq \mu_e(M)
\end{aligned}$$

(iii) $\mu_e(M \cap A) = \Sigma_{n=1}^{\infty} \mu_e(M \cap A_n)$ for each $M \subset X$.

(12-7)$^+$ Show that

(i) $(X, \mathfrak{M}_e, \mu_e)$ is a measure space. Hint: Use (12-5) and (12-6).

(ii) μ_e is a complete measure on $\mathfrak{M}_e$.

(iii) Suppose that for each $M \subset X$ with $\mu_e(M) < +\infty$ there is a $B \in \mathfrak{M}_e$ such that $M \subset B$ and $\mu_e(B) < +\infty$. Then μ_e is a saturated measure on $\mathfrak{M}_e$. Hint: If $A \cap B \in \mathfrak{M}_e$ and $M \subset B$, then

$$\mu_e(M) = \mu_e[M \cap (A \cap B)] + \mu_e[M - (A \cap B)]$$
$$= \mu_e(M \cap A) + \mu_e(M - A)$$

The measure space $(X, \mathfrak{M}_e, \mu_e)$ is said to be induced by the outer measure μ_e. We note that for the sake of brevity we have used the symbol μ_e to denote both the outer measure and its restriction to $\mathfrak{M}_e$.

(12-8) Let X be a set, and let $\|X\| \geq \aleph_2$. For $A \subset X$, set $\mu_e(A) = 0$ if $\|A\| \leq \aleph_0$, $\mu_e(A) = 1$ if $\|A\| = \aleph_1$, and $\mu_e(A) = +\infty$ if $\|A\| \geq \aleph_2$. Show that

(i) μ_e is an outer measure on X.

(ii) A set $A \subset X$ is μ_e-measurable iff A or $X - A$ is countable.

(iii) The measure μ_e on $\mathfrak{M}_e$ is not saturated.

(12-9)+ Let $(X, \mathfrak{M}, \mu)$ be a measure space. For $A \subset X$, set

$$\mu_e(A) = \inf\{\mu(B) : B \in \mathfrak{M}, A \subset B\}$$

Show that

(i) If $A_n \subset X$, $n = 1, 2, \ldots$, then

$$\mu_e\left(\bigcup_{n=1}^{\infty} A_n\right) \leq \sum_{n=1}^{\infty} \mu_e(A_n)$$

Hint: Assume $\mu_e(A_n) < +\infty$, $n = 1, 2, \ldots$, choose $\epsilon > 0$, and select $B_n \in \mathfrak{M}$ so that $A_n \subset B_n$ and $\mu(B_n) < \mu_e(A_n) + \epsilon 2^{-n}$. Use (8.4).

(ii) μ_e is an outer measure on X. This outer measure is said to be induced by the measure μ.

In exercises (12-10) through (12-12) we shall assume that $(X, \mathfrak{M}, \mu)$ is a measure space and that μ_e is the outer measure induced by μ. By $\mathfrak{M}_e$ we shall denote the σ-algebra of all μ_e-measurable subsets of X.

(12-10)$^+$ Show that

(i) If $M \subset X$, then there is a $B \in \mathfrak{M}$ such that $M \subset B$ and $\mu_e(M) = \mu(B)$.

(ii) $\mathfrak{M} \subset \mathfrak{M}_e$ and $\mu_e(A) = \mu(A)$ for each $A \in \mathfrak{M}$. Hint: Given $M \subset X$, use (i) to choose a $B \in \mathfrak{M}$ so that $M \subset B$ and $\mu_e(M) = \mu(B)$. Observe that for each $A \in \mathfrak{M}$,

$$\begin{aligned}\mu_e(M) &= \mu_e(B) = \mu(B \cap A) + \mu(B - A) \\ &\geq \mu_e(M \cap A) + \mu_e(M - A) \geq \mu_e(M)\end{aligned}$$

(iii) μ_e is a complete and saturated measure on $\mathfrak{M}_e$. Hint: Use (i), (ii), and (12-7)(iii).

(12-11)$^+$ Show that

(i) If $M \in \mathfrak{M}_e$ and $\mu_e(M) < +\infty$, then there are $A, B \in \mathfrak{M}$ such that $A \subset M \subset B$, $\mu(B) < +\infty$, and $\mu(B - A) = 0$. Hint: Use (12-10)(i) to choose $B \in \mathfrak{M}$ so that $M \subset B$ and $\mu_e(M) = \mu(B)$. Then $\mu_e(B - M) = 0$. Using (12-10)(i) again, choose $C \in \mathfrak{M}$ so that $B - M \subset C$ and $\mu(C) = 0$. It suffices to let $A = B - C$.

(ii) If $M \in \mathfrak{M}_e$ is μ_e-σ-finite, then there are $A, B \in \mathfrak{M}$ such that $A \subset M \subset B$, B is μ-σ-finite, and $\mu(B - A) = 0$. Hint: Mimic the proof of (12.13).

(iii) Let $\bar{\mu}$ on $\bar{\mathfrak{M}}$ be the completion of the measure μ on $\mathfrak{M}$ [see (8-9)]. Then $\bar{\mathfrak{M}} \subset \mathfrak{M}_e$ and $\mu_e(M) = \bar{\mu}(M)$ for each $M \in \bar{\mathfrak{M}}$. Hint: Use (12-10)(ii), (12-7)(ii), and (8-9)(v).

(iv) If $M \in \mathfrak{M}_e$ is μ_e-σ-finite, then $M \in \bar{\mathfrak{M}}$. Hint: Use (ii).

(v) Let $(\bar{\mu})^\wedge$ on $(\bar{\mathfrak{M}})^\wedge$ be the saturation of the measure $\bar{\mu}$ on $\bar{\mathfrak{M}}$ [see (8-10)]. Then $(\bar{\mathfrak{M}})^\wedge \subset \mathfrak{M}_e$ and $\mu_e(M) = (\bar{\mu})^\wedge(M)$ for each $M \in (\bar{\mathfrak{M}})^\wedge$. Hint: Use (iii), (iv), and (12-10)(i) and (iii).

(vi) $\mathfrak{M}_e = (\bar{\mathfrak{M}})^\wedge$. Hint: Let $M \in \mathfrak{M}_e$ and $A \in \bar{\mathfrak{M}}$ such that $\bar{\mu}(A) < +\infty$. Observe that $M \cap A \in \mathfrak{M}_e$ and $\mu_e(M \cap A) < +\infty$, and apply (iv).

Thus the measure μ_e on $\mathfrak{M}_e$ is the completion and saturation [in this order: see (11-3)] of the measure μ on $\mathfrak{M}$.

(12-12)$^+$ Let I be the integral induced by μ, and let $(X, \mathfrak{I}, \iota)$ be the measure space induced by I. Prove

(i) $A \in \mathfrak{M}_e$ and $\mu_e(A) < +\infty$ iff $A \in \mathfrak{I}$ and $\iota(A) < +\infty$. In either case $\mu_e(A) = \iota(A)$. Hint: Use (12-10)(ii), (12-11)(i), (12.12), and the fact that both μ_e and ι are complete.

(ii) $(X, \mathfrak{M}_e, \mu_e) = (X, \mathfrak{I}, \iota)$. Hint: Use (i) and the fact that both μ_e and ι are saturated.

Combining (ii) with (12-11)(vi), we obtain a very pleasing description of $(X, \mathfrak{I}, \iota)$: The measure space $(X, \mathfrak{I}, \iota)$ is obtained from the measure space $(X, \mathfrak{M}, \mu)$ by the completion and saturation (in this order) of the measure μ.

Prove this result directly without using the outer measure μ_e.

13. INTEGRALS INDUCED BY MEASURES IN A TOPOLOGICAL SPACE

If X is a topological space, we shall use the results of Chapter 12 to establish a one-to-one correspondence between nonnegative linear functionals on $C_0(X)$ and regular measures in X.

Throughout this chapter we shall assume the following:

(i) X is a locally compact Hausdorff space

(ii) $\mathfrak{M}$ is a σ-algebra in X and μ is a measure on $\mathfrak{M}$

(iii) I and $(X, \mathfrak{I}, \iota)$ are, respectively, the integral and the measure space induced by the measure μ [see (12.10)].

The reader should keep in mind that according to Theorem (12.17), $(X, \mathfrak{I}, \iota)$ is equal to $(X, \mathfrak{M}, \mu)$ whenever the measure μ is complete and saturated.

By $\mathfrak{Z}$ and $\mathfrak{G}$ we shall denote the families of all compact and all open subsets of X, respectively. The family of all σ-compact subsets of X is denoted by $\mathfrak{Z}_\sigma$ and the family of all G_δ subsets of X is denoted by $\mathfrak{G}_\delta$.

(13.1) Proposition. If μ is a regular measure, so is ι.

Proof. By (12.14), it suffices to show that

$$\iota(A) = \inf\{\iota(G) : G \in \mathfrak{G},\ A \subset G\}$$

for each $A \in \mathfrak{I}$ [see (9.1)]. As this holds trivially if $\iota(A) = +\infty$, we choose $A \in \mathfrak{I}$ with $\iota(A) < +\infty$. It follows from (12.12), (12.14), and (10.18) that there is a $C \in \mathfrak{M}$ such that $A \subset C$ and $\iota(A) = \mu(C)$. From the regularity of μ and (12.14) we obtain

$$\iota(A) = \mu(C) = \inf\{\mu(G) : G \in \mathfrak{G},\ C \subset G\} \geq \inf\{\mu(G) : G \in \mathfrak{G},\ A \subset G\}$$
$$= \inf\{\iota(G) : G \in \mathfrak{G},\ A \subset G\} \geq \iota(A) \quad ///$$

Exercise (13-1) will show that the converse of Proposition (13.1) is generally false.

When X is a topological space, a natural question is under what conditions $\mathfrak{C}_o(X) \subset \mathfrak{L}$. In order to answer this question we shall need few lemmas.

(13.2) Lemma. Let $f \in \mathfrak{C}_o(X)$ and let $0 < a < +\infty$. If

$$F = \{x \in X : f(x) \geq a\}$$

then $F \in \mathfrak{F} \cap \mathfrak{G}_\delta$.

Proof. Since F is a closed subset of supp f, it is compact. On the other hand,

$$F = \bigcap_{n=1}^{\infty} \{x \in X : f(x) > a - \frac{1}{n}\}$$

and so F is G_δ. ///

(13.3) Corollary. Let $F \in \mathfrak{F}$, $G \in \mathfrak{G}$, and $F \subset G$. Then there is a $C \in \mathfrak{F} \cap \mathfrak{G}_\delta$ such that $F \subset C \subset G$.

Proof. Choose $G_1 \in \mathfrak{G}$ so that $G_1^- \in \mathfrak{F}$ and $F \subset G_1 \subset G$. Then there is an $f \in \mathfrak{C}(X)$ for which $0 \leq f \leq 1$, $f(x) = 1$ if $x \in F$, and $f(x) = 0$ if $x \in X - G_1$. By (13.2), it suffices to let

$$C = \{x \in X : f(x) \geq \frac{1}{2}\} \quad ///$$

(13.4) Lemma. Let $F \in \mathfrak{F} \cap \mathfrak{G}_\delta$. Then there is a nonnegative function $f \in \mathfrak{C}(X)$ such that

$$F = \{x \in X : f(x) = 0\}$$

Proof. Choose $G_n \in \mathfrak{G}$, $n = 1, 2, \ldots$, so that $F = \bigcap_{n=1}^{\infty} G_n$. Then there are $f_n \in \mathfrak{C}(X)$ for which $0 \leq f_n \leq 2^{-n}$, $f_n(x) = 0$ if $x \in F$, and $f_n(x) = 2^{-n}$ if $x \in X - G_n$. Let $f = \Sigma_{n=1}^{\infty} f_n$. Since the series $\Sigma_{n=1}^{\infty} f_n$ converges uniformly, we have $f \in \mathfrak{C}(X)$. Clearly, $f(x) = 0$ for each $x \in F$. If $x \in X - F$ then $x \in X - G_n$ for some integer $n \geq 1$, and so

$$f(x) \geq f_n(x) = 2^{-n} > 0 \quad ///$$

(13.5) Corollary. Let $F \in \mathfrak{J} \cap \mathfrak{G}_\delta$. Then there is a nonnegative function $f \in C_o(X)$ such that

$$F = \{x \in X : f(x) = 1\}$$

Proof. Choose $G \in \mathfrak{G}$ so that $G^- \in \mathfrak{J}$ and $F \subset G$. Then there is a $g \in C(X)$ for which $0 \leq g \leq 1$, $g(x) = 1$ if $x \in F$, and $g(x) = 0$ if $x \in X - G$. By (13.4), there is an $h \in C(X)$ such that $h \geq 0$ and

$$F = \{x \in X : h(x) = 0\}$$

Clearly, $f = g(1 - h)^+$ is the required function. ///

(13.6) Proposition. $C_o(X) \subset \mathfrak{L}$ iff $\mathfrak{J} \cap \mathfrak{G}_\delta \subset \mathfrak{J}$ and $\iota(F) < +\infty$ for each $F \in \mathfrak{J} \cap \mathfrak{G}_\delta$.

Proof. Let $\mathfrak{J} \cap \mathfrak{G}_\delta \subset \mathfrak{J}$ and $\iota(F) < +\infty$ for each $F \in \mathfrak{J} \cap \mathfrak{G}_\delta$. Choose $f \in C_o(X)$. It follows from (13.2) that f^+ and f^- are $\mathfrak{J}$-measurable; for if $a \leq 0$, then

$$\{x \in X : f^+(x) \geq a\} = \{x \in X : f^-(x) \geq a\} = X$$

Let $\alpha = \sup\{|f(x)| : x \in X\}$ and choose $F \in \mathfrak{J} \cap \mathfrak{G}_\delta$ so that supp $f \subset F$ [see (13.3)]. Then $|f| \leq \alpha\chi_F$ and it follows from (10.5), (6.15), and (6.12) that $|f| \in \mathfrak{L}$. By (7.14), (7.13), and (10.16) also $f \in \mathfrak{L}$.

Conversely, let $C_o(X) \subset \mathfrak{L}$, and let $F \in \mathfrak{J} \cap \mathfrak{G}_\delta$. By (13.5) there is a nonnegative function $f \in C_o(X)$ such that

$$F = \{x \in X : f(x) = 1\}$$

By (10.13), $F \in \mathfrak{J}$, and since $\chi_F \leq f$, it follows that

$$\iota(F) = I\chi_F \leq If < +\infty \quad ///$$

(13.7) Corollary. If ι is a regular measure, then $C_o(X) \subset \mathfrak{L}$.

Recall that $\mathfrak{L}_\mu$ denotes the system of all functions f on X for which a finite $\int f\, d\mu$ exists.

(13.8) Corollary. If μ is a regular measure, then $C_o(X) \subset \mathfrak{L}_\mu$.

Proof. Since every continuous function on X is measurable with respect to the Borel σ-algebra in X, the corollary follows from (13.1), (13.7), and (12.19). ///

The following theorem has many important applications in various branches of mathematics. In the literature it is usually referred to as the "Riesz representation theorem."

(13.9) Theorem (Riesz). Let J be a nonnegative linear functional on $\mathfrak{C}_o(X)$. Then there is a unique measure space $(X, \mathfrak{M}, \nu)$ satisfying the following conditions:

(i) ν is a regular complete and saturated measure

(ii) $Jf = \int f\, d\nu$ for each $f \in \mathfrak{C}_o(X)$.

Proof. By (4.7), J is a Bourbaki integral on $\mathfrak{C}_o(X)$. Thus it follows from (10.18), (10.19), and (12.22) that the measure space $(X, \mathfrak{M}, \nu)$ induced by J satisfies conditions (i) and (ii) of the theorem.

Let $(X, \mathfrak{M}_i, \nu_i)$, $i = 1, 2$, be any measure spaces which satisfy conditions (i) and (ii) of the theorem and let $F \in \mathfrak{F}$. Since both ν_1 and ν_2 are regular, it is easy to construct a decreasing sequence $\{G_n\} \subset \mathfrak{G}$ so that

$$F \subset G_n \qquad G_n^- \in \mathfrak{F}$$

$n = 1, 2, \ldots$, and $\lim \nu_i(G_n) = \nu_i(F)$, $i = 1, 2$. Find $g_n \in \mathfrak{C}(X)$ such that $0 \le g_n \le 1$, $g_n(x) = 1$ if $x \in F$, and $g_n(x) = 0$ if $x \in X - G_n$. Setting $f_n = \wedge_{k=1}^{n} g_k$, we see that $f_n \in \mathfrak{C}_o(X)$ and $f_n(x) \searrow \chi_F(x)$ for each $x \in X - (\cap_{n=1}^{\infty} G_n - F)$. Because $\nu_i(\cap_{n=1}^{\infty} G_n - F) = \lim \nu_i(G_n) - \nu(F) = 0$ $f_n \searrow \chi_F$ ν_i-almost everywhere, $i = 1, 2$. By (12.19) and (12.21), the abstract Lebesgue integral with respect to ν_i is equal to the integral I_i induced by ν_i, $i = 1, 2$. By (12.17), the measure space induced by I_i is equal to $(X, \mathfrak{M}_i, \nu_i)$, $i = 1, 2$. Thus it follows from the dual of (11.3) that

$$\nu_1(F) = \int \chi_F d\nu_1 = \lim \int f_n d\nu_1 = \lim Jf_n$$
$$= \lim \int f_n d\nu_2 = \int \chi_F d\nu_2 = \nu_2(F)$$

An application of (9.10) gives $(X, \mathfrak{N}_1, \nu_1) = (X, \mathfrak{N}_2, \nu_2)$. ///

(13.10) Remark. By (13.8), condition (i) of Theorem (13.9) implies that $\int f\, d\nu$ exists for each $f \in C_0(X)$. Therefore there is no ambiguity in the formulation of condition (ii).

If $C_0(X) \subset \mathfrak{L}$ we shall denote by J the integral I restricted to $C_0(X)$. According to (4.7), J is a Bourbaki integral on $C_0(X)$. Since $C_0(X)$ is a Stonian system, the integral J induces a measure space which we denote by $(X, \mathfrak{N}, \nu)$. We shall adhere to this notation throughout the remainder of this chapter.

(13.11) Definition. If $C_0(X) \subset \mathfrak{L}$, then the integral J and the measure space $(X, \mathfrak{N}, \nu)$ are said to be topologically induced by the measure μ.

It is an immediate consequence of Theorem (12.18) that the measure space $(X, \mathfrak{N}, \nu)$ is topologically induced also by the measure ι.

(13.12) Proposition. If $C_0(X) \subset \mathfrak{L}$, then $(X, \mathfrak{N}, \nu)$ is the unique measure space satisfying the following conditions:

(i) ν is a regular complete and saturated measure

(ii) $\int f\, d\nu = \int f\, d\iota$ for each $f \in C_0(X)$.

Proof. The proposition follows directly from (10.18), (10.19), (12.22), and (13.9). ///

(13.13) Corollary. Let ι be a regular measure. Then $(X, \mathfrak{N}, \nu) = (X, \mathfrak{J}, \iota)$.

Proof. By (13.7), $C_0(X) \subset \mathfrak{L}$. Since ι is complete and saturated [see (10.18)], the corollary follows from (13.12). ///

In general, there is little connection between the measures ν and μ, or ν and ι [see (13-2)]. However, ν and ι are closely related whenever $\mathfrak{G} \subset \mathfrak{J}_\sigma$, i.e., whenever each open subset of X is σ-compact.

The following observation is immediate.

(13.14) If $\mathfrak{G} \subset \mathfrak{J}_\sigma$, then $\mathfrak{J} \subset \mathfrak{G}_\delta$.

We note that the converse of (13.14) is generally false [see (13-4) (c)].

(13.15) Proposition. Suppose that $\mathfrak{G} \subset \mathfrak{K}_\sigma$. Let $\mathfrak{K} \subset \mathfrak{J}$ and $\iota(F) < +\infty$ for each $F \in \mathfrak{K}$. Then $\mathfrak{N} \subset \mathfrak{J}$ and $\nu(A) = \iota(A)$ for each $A \in \mathfrak{N}$.

Proof. According to (13.6) the integral J and the measure space $(X, \mathfrak{N}, \nu)$ are defined. For the purpose of this proof only we shall use the symbols $\mathfrak{F}$, $\mathfrak{F}_+$, $\mathfrak{L}$, $\mathfrak{L}_+$, etc., with the subscripts I and J, according to whether they refer to the integrals I and J, respectively. It follows from (3.13) that the Daniell and Bourbaki extensions of $\mathfrak{F}_J = C_0(X)$ coincide. Because $\mathfrak{F}_J \subset \mathfrak{L}_I$ [see (13.6)], it follows from (6.17) and its dual that $\mathfrak{F}_{J+} \subset \mathfrak{L}_{I+}$, $\mathfrak{F}_{J-} \subset \mathfrak{L}_{I-}$, and $Jf = If$ for each $f \in \bar{\mathfrak{F}}_J$. Therefore,

$$\underline{J}f \le \underline{I}f \le \bar{I}f \le \bar{J}f$$

for each function f on X. This implies that $\mathfrak{L}_J \subset \mathfrak{L}_I$ and $Jf = If$ for each $f \in \mathfrak{L}_J$. Because X is σ-compact, it is easy to see that the function identically equal to $+\infty$ on X belongs to $\mathfrak{F}^*_{J+}$. From (6.19) and (6.17) we obtain that $\mathfrak{L}_{J+} \subset \mathfrak{L}_{I+}$ and $Jf = If$ for each $f \in \mathfrak{L}_{J+}$. The proposition follows. ///

It is noteworthy to mention that the inclusion $\mathfrak{N} \subset \mathfrak{J}$ may be proper in the previous proposition [see (13-3)].

(13.16) Corollary. Let $\mathfrak{M}$ be equal to the Borel σ-algebra $\mathfrak{B}$ in X, and let $\mu(F) < +\infty$ for each $F \in \mathfrak{K}$. If $\mathfrak{G} \subset \mathfrak{K}_\sigma$, then the measure μ is regular.

Proof. By (12.14), $\mathfrak{M} \subseteq \mathfrak{J}$ and $\mu(A) = \iota(A)$ for each $A \in \mathfrak{M}$. In particular, $\mathfrak{K} \subset \mathfrak{J}$ and $\iota(F) < +\infty$ for each $F \in \mathfrak{K}$. Thus by (13.15), $\mathfrak{N} \subset \mathfrak{J}$ and $\nu(A) = \iota(A)$ for each $A \in \mathfrak{N}$. Since ν is regular [see (13.12)(i)] and since $\mathfrak{M} = \mathfrak{B}$, we have $\mathfrak{M} \subset \mathfrak{N}$ and

$$\mu(A) = \iota(A) = \nu(A)$$

for each $A \in \mathfrak{M}$. The corollary follows from the regularity of ν. ///

A substantial generalization of this corollary will be given in Chapter 18 [see (18.31)].

Exercises

In all exercises X will be a topological space. We shall use the following notation:

If $(X, \mathfrak{M}, \mu)$ is a measure space, we denote by I and $(X, \mathfrak{J}, \iota)$, respectively, the integral and the measure space induced by the measure μ [see (12.10)]. If, in addition, X is a locally compact Hausdorff space and each function from $C_o(X)$ is I-summable, we denote by J and $(X, \mathfrak{N}, \nu)$, respectively, the integral and the measure space topologically induced by the measure μ [see (13.11)].

(13-1) Let $X = \mathbf{R}$, and let Q denote the set of all rational numbers. Consider the measure space $(X, \mathfrak{M}, \mu)$, where $\mathfrak{M}$ consists of all subsets of Q and their complements in $\mathbf{R}$, and μ is the Dirac measure at $x_o \in Q$ on $\mathfrak{M}$ [see (8-1)(v)]. Show that

(i) $\mathfrak{J} \cap \mathfrak{G}_\delta \not\subset \mathfrak{M}$; in particular, μ is not a regular measure.

(ii) $\mathfrak{J} = \exp X$ and ι is the Dirac measure at x_o on $\mathfrak{J}$.

(iii) $\mathfrak{N} = \mathfrak{J}$ and $\nu = \iota$.

(13-2) Let $(X, \mathfrak{M}, \mu)$ be the measure space from (9-10). Show that

(i) $\mathfrak{J} = \mathfrak{M}$ and $\iota = \mu$.

(ii) $\mathfrak{N} = \exp X$ and ν is the Dirac measure at Ω on $\mathfrak{N}$ [see (8-1)(v)].

(13-3) Let $(X, \mathfrak{M}, \mu)$ be a measure space such that $X = \mathbf{R}$, $\Lambda \subset \mathfrak{M}$, and $\mu(A) = \lambda(A)$ for each $A \in \Lambda$ [see (10-7)]. Show that

(i) $\mathfrak{J} = \bar{\mathfrak{M}}$ and $\iota = \bar{\mu}$ [see (8-9)]. Hint: Use (12-12) and (8.16).

(ii) $\mathfrak{N} = \Lambda$ and $\nu = \lambda$. Hint: Observe that the measures topologically induced by μ and λ coincide and use (13.13).

Note: In Ref. 15, there is a proof of the existence of a measure space $(\mathbf{R}, \mathfrak{M}, \mu)$ such that $\Lambda \subsetneq \mathfrak{M}$ and $\mu(A) = \lambda(A)$ for each $A \in \Lambda$.

(13-4)* Let X be a locally compact Hausdorff space. Prove that

(i) If $X \in \mathfrak{J}_\sigma$, then there are $X_n \in \mathfrak{J}$ such that $X_n \subset X^o_{n+1}$, $n = 1, 2, \ldots$, and $X = \bigcup_{n=1}^\infty X_n$.

(ii) Let $X \in \mathfrak{J}_\sigma$ and $\mathfrak{J} \subset \mathfrak{G}_\delta$. Then each closed set $F \subset X$ is G_δ. Hint: Choose X_n, $n = 1, 2, \ldots$, as in (i), and let $X_0 = X_{-1} = \emptyset$.

For n = 1, 2, ..., set $F_n = F \cap (X_n - X^o_{n-1})$ and take $G_{n,k} \in \mathfrak{G}$ so that $G_{n,k} \subset X_{n+1} - X_{n-2}$, k = 1, 2, ..., and $G_{n,k} \searrow F_n$ as $k \to +\infty$. Observe that $F = \bigcap_{k=1}^{\infty} G_k$ where $G_k = \bigcup_{n=1}^{\infty} G_{n,k}$.

(iii) Let X be paracompact and $\mathfrak{F} \subset \mathfrak{G}_\delta$. Then each closed set $F \subset X$ is G_δ. <u>Hint</u>: Use (ii) and the fact that every paracompact locally compact Hausdorff space is a free union of σ-compact spaces (see Ref. 8, Chapter XI, Theorem 7.3, p. 241).

(iv) Let $X \in \mathfrak{F}_\sigma$. Then $\mathfrak{G} \subset \mathfrak{F}_\sigma$ iff $\mathfrak{F} \subset \mathfrak{G}_\delta$. <u>Hint</u>: Observe that $G \in \mathfrak{F}_\sigma$ iff G is F_σ; apply (ii).

Show by example that

(a) Assertion (i) is false if X is not locally compact. <u>Hint</u>: Use the rationals.

(b) Assertion (iii) is false if X is not paracompact. <u>Hint</u>: Use W and the closed subset of all limit ordinals.

(c) Assertion (iv) is false if X is not σ-compact. <u>Hint</u>: Use an uncountable discrete space.

<u>(13-5)</u>$^+$ Let $(X, \mathfrak{M}, \mu)$ be a measure space where X is a locally compact Hausdorff space and μ is a regular measure. Show that supp μ = supp J [see (4-14) and (9-8)].

<u>(13-6)</u> Use (12-12), (9-5), and (9-6) to give an alternate proof of Proposition (13.1).

The remaining exercises will be concerned with an important construction of regular measures by means of outer measures [see (12-3) - (12-12)]. Throughout we shall assume that X is a locally compact Hausdorff space.

A family $\mathfrak{F}_o \subset \mathfrak{F}$ is called a <u>sufficient family of compact sets</u> in X if

(a) $\emptyset \in \mathfrak{F}_o$

(b) $C, D \in \mathfrak{F}_o \Rightarrow C \cup D \in \mathfrak{F}_o$

(c) For each $F \in \mathfrak{F}$ and $G \in \mathfrak{G}$ with $F \subset G$ there is a $C \in \mathfrak{F}_o$ such that $F \subset C \subset G$.

Let $\mathfrak{F}_o$ be a sufficient family of compact sets. A nonnegative real-valued function v on $\mathfrak{F}_o$ is called a <u>volume</u> on $\mathfrak{F}_o$ whenever (α) $v(C) \leq v(D)$ for each $C, D \in \mathfrak{J}_o$ for which $C \subset D$

(β) $v(C \cup D) \leq v(C) + v(D)$ for each $C, D \in \mathfrak{J}_o$

(γ) $v(C \cup D) = v(C) + v(D)$ for each disjoint $C, D \in \mathfrak{J}_o$.

A volume space is a triple $(X, \mathfrak{J}_o, v)$, where $\mathfrak{J}_o$ is a sufficient family of compact sets in X and v is a volume on $\mathfrak{J}_o$.

In exercises (13-7) through (13-10) we shall assume that $(X, \mathfrak{J}_o, v)$ is a volume space.

(13-7)*+ For $G \in \mathfrak{G}$, let

$$w(G) = \sup\{v(C) : C \in \mathfrak{J}_o, C \subset G\}$$

and show that

(i) $w(C) = v(C)$ for each $C \in \mathfrak{G} \cap \mathfrak{J}_o$.

(ii) $w(\emptyset) = 0$. Hint: Use (γ) to show that $v(\emptyset) = 0$.

(iii) $w(G) \leq w(H)$ for each $G, H \in \mathfrak{G}$ for which $G \subset H$.

(iv) $w(G \cup H) \leq w(G) + w(H)$ for each $G, H \in \mathfrak{G}$. Hint: Let $C \in \mathfrak{J}_o$ and $C \subset G \cup H$. Use (c) to show that there are $C_1, C_2 \in \mathfrak{J}_o$ such that $C_1 \subset G$, $C_2 \subset H$, and $C \subset C_1 \cup C_2$.

(v) $w(\bigcup_{n=1}^{\infty} G_n) \leq \Sigma_{n=1}^{\infty} w(G_n)$ for each sequence $\{G_n\} \subset \mathfrak{G}$. Hint: By induction extend (iv) to finite unions. Observe that if $C \in \mathfrak{J}_o$ and $C \subset \bigcup_{n=1}^{\infty} G_n$, then $C \subset \bigcup_{n=1}^{N} G_n$ for some integer $N \geq 1$.

(vi) $w(G \cup H) = w(G) + w(H)$ for each disjoint $G, H \in \mathfrak{G}$.

(vii) $w(\bigcup_{n=1}^{\infty} G_n) = \Sigma_{n=1}^{\infty} w(G_n)$ for each disjoint sequence $\{G_n\} \subset \mathfrak{G}$. Hint: By induction extend (vi) to finite disjoint unions and use (iii) and (v).

The function $G \mapsto w(G)$, $G \in \mathfrak{G}$, is called the inner (or interior) volume on $\mathfrak{G}$ induced by the volume v.

(13-8)+ Let w be the inner volume on $\mathfrak{G}$ induced by v. For $A \subset X$ set

$$\mu_v(A) = \inf\{w(G) : G \in \mathfrak{G}, A \subset G\}$$

and show that

(i) $\mu_v(G) = w(G)$ for each $G \in \mathfrak{G}$.

(ii) If $A \subset B \subset X$ then $\mu_v(A) \leq \mu_v(B)$.

(iii) If $\{A_n\}$ is a countable family of subsets of X, then

$$\mu_v(\bigcup_n A_n) \leq \sum_n \mu_v(A_n)$$

Hint: Assume that $\mu_v(A_n) < +\infty$ for each n. Given $\epsilon > 0$, choose $G_n \in \mathfrak{G}$ so that $A_n \subset G_n$ and $w(G_n) < \mu_v(A_n) + \epsilon 2^{-n}$, and use (13-7)(v).

Conclude that μ_v is an outer measure on X. It is said to be induced by the volume v.

(13-9)$^+$ Let μ_v be the outer measure induced by the volume v. Show that

(i) $\mu_v(C^o) \leq v(C) \leq \mu_v(C)$ for each $C \in \mathfrak{I}_o$.

(ii) For each $C \in \mathfrak{I}_o$, let

$$v(C) = \inf\{v(D) : D \in \mathfrak{I}_o \text{ , } C \subset D^o\}$$

Then $\mu_v(C) = v(C)$ for each $C \in \mathfrak{I}_o$. Hint: Use (i).

(13-10)$^{*+}$ Let μ_v be the outer measure induced by the volume v, and let $\mathfrak{M}_v$ denote the σ-algebra of all μ_v-measurable subsets of X. Prove

(i) $A \in \mathfrak{M}_v$ iff

$$\mu_v(G) \geq \mu_v(G \cap A) + \mu_v(G - A)$$

for each $G \in \mathfrak{G}$.

(ii) $\mathfrak{G} \subset \mathfrak{M}_v$. Hint: Let $H \in \mathfrak{G}$, and choose $G \in \mathfrak{G}$ with $\mu_v(G) < +\infty$. Given $\epsilon > 0$, find $C, D \in \mathfrak{I}_o$ so that $C \subset G \cap H$, $D \subset G - C$, and

$$\mu_v(G \cap H) < v(C) + \epsilon \qquad \mu_v(G - C) < v(D) + \epsilon$$

Observe that

$$\mu_v(G \cap H) + \mu_v(G - H) < v(C) + v(D) + 2\epsilon$$
$$= v(C \cup D) + 2\epsilon \leq v(G) + 2\epsilon$$

(iii) μ_v on $\mathfrak{M}_v$ is a regular, complete, and saturated measure.
Hint: Use (12-7)(iii).

The measure space $(X, \mathfrak{M}_v, \mu_v)$ is said to be induced by the volume v.

(13-11)$^{*+}$ Let $(X, \mathfrak{M}, \mu)$ be a measure space, $\mathfrak{I} \cap \mathfrak{G}_\delta \subset \mathfrak{M}$, and let $\mu(C) < +$

for each $C \in \mathfrak{J} \cap \mathfrak{G}_\delta$. Set $\mathfrak{J}_o = \mathfrak{J} \cap \mathfrak{G}_\delta$ and $v(C) = \mu(C)$ for each $C \in \mathfrak{J}_o$. Show that $(X, \mathfrak{J}_o, v)$ is a volume space. Hint: Use (13.3).

Denote by $(X, \mathfrak{M}_v, \mu_v)$ the measure space induced by the volume v and prove the following:

(i) If μ is a regular measure, then $\mathfrak{M} \subset \mathfrak{M}_v$ and $\mu_v(A) = \mu(A)$ for each $A \in \mathfrak{M}$. Hint: Complete and saturate the measure μ. Then use (13-9)(ii), (13-10)(iii), and (9-11).

(ii) If μ is a regular, complete, and saturated measure, then $(X, \mathfrak{M}_v, \mu_v) = (X, \mathfrak{M}, \mu)$. Hint: Use (13-9)(ii), (13-10)(iii), and (9-11).

(iii) If μ is a regular measure, then μ_v is the outer measure induced by μ [see (12-9)].

(iv) $\mu_v(G) = \nu(G)$ for each $G \in \mathfrak{G}$. Hint: Since χ_G is a lower semicontinuous function [see (3.3)],

$$\nu(G) = \sup\{\mathrm{J}f : f \in C_o(X),\ 0 \leq f \leq \chi_G\}$$

(v) $(X, \mathfrak{M}_v, \mu_v) = (X, \mathfrak{N}, \nu)$. Hint: Use (iv), (3-10)(iii), and (9.10).

Applying (iii), derive (i) and (ii) from (12-10)(ii) and (12-11)(v), (vi), respectively.

(13-12)$^+$ Let $(X, \mathfrak{M}, \mu)$ be a measure space, and suppose that every function from $C_o(X)$ is I-summable. Let $\mathfrak{J}_o = \mathfrak{J} \cap \mathfrak{G}_\delta$ and $v(C) = I\chi_C$ for each $C \in \mathfrak{J}_o$. Show that

(i) $(X, \mathfrak{J}_o, v)$ is a volume space. Hint: Use (13.3).

(ii) If $(X, \mathfrak{M}_v, \mu_v)$ is the measure space induced by the volume v, then $(X, \mathfrak{M}_v, \mu_v) = (X, \mathfrak{N}, \nu)$. Hint: Observe that $(X, \mathfrak{N}, \nu)$ is topologically induced by ι, and apply (13-11)(v).

(13-13) Let $(X, \mathfrak{M}, \mu)$ be the measure space from (9-10). Set $\mathfrak{J}_o = \mathfrak{J}$ and $v(C) = \mu(C)$ for each $C \in \mathfrak{J}_o$. Show that

(i) $(X, \mathfrak{J}_o, v)$ is a volume space.

(ii) If $(X, \mathfrak{M}_v, \mu_v)$ is the measure space induced by the volume v, then $\mathfrak{M}_v = \exp X$ and μ_v is the Dirac measure at Ω.

(13-14)$^+$ Let $X = \mathbf{R}$, and let g be a nondecreasing real-valued function on X. Denote by $\mathfrak{J}_o$ the family of all finite unions (possibly empty) of disjoint compact intervals. If $C \in \mathfrak{J}_o$ and $C = \bigcup_{i=1}^{n} [a_i, b_i]$ with $a_1 < b_1 < \cdots < a_n < b_n$, set

$$v_g(C) = \sum_{i=1}^{n} [g(b_i) - g(a_i)]$$

Show that

(i) $(X, \mathfrak{J}_o, v_g)$ is a volume space.

(ii) v_g induces the measure space $(\mathbf{R}, \Lambda_g, \lambda_g)$ from (10-9).

14. INTEGRATION OVER MEASURABLE SETS

So far we have been integrating functions defined on the whole space X. In this chapter we shall show how to integrate functions defined only on a measurable subset of X.

The idea is quite simple. If f is an integrable function on X and $A \subset X$ is a measurable set, then by (10.10), $f\chi_A$ is also an integrable function on X. Rather naturally, the integral of $f\chi_A$ over X is called an integral of f over A. This is, indeed, a good definition, for the values of $f\chi_A$ depend only on the values which f takes on A.

Throughout this chapter we shall assume that $(X, \mathfrak{M}, \mu)$ is an arbitrary measure space and we shall denote by $\mathfrak{m}(X)$ the family of all measurable functions on X. By $\bar{\mathcal{L}}(X)$ we shall denote the family of all functions $f \in \mathfrak{m}(X)$ for which the abstract Lebesgue integral $\int f\, d\mu$ exists [see (8.17)], and we let

$$\mathcal{L}(X) = \{f \in \bar{\mathcal{L}}(X) : \int f\, d\mu \neq \pm\infty\}$$

Recall that by (12.19) the abstract Lebesgue integral is a restriction of a suitably defined Daniell integral.

If $A \subset X$ and f is a function on A we shall define a function $f^\wedge$ on X as follows:

$$f^\wedge(x) = \begin{cases} f(x) & \text{if } x \in A \\ 0 & \text{if } x \in X - A \end{cases}$$

The function $f^\wedge$ is called the <u>normal extension</u> of f.

<u>(14.1)</u> <u>Definition</u>. Let f be a function on a set $A \in \mathfrak{M}$, and let $f^\wedge$ be its normal extension. We say that f is <u>measurable</u>, or <u>integrable</u>, or <u>summable on A</u> whenever $f^\wedge \in \mathfrak{m}(X)$, or $f^\wedge \in \bar{\mathcal{L}}(X)$, or $f^\wedge \in \mathcal{L}(X)$, respectively.

The families of all measurable, integrable, and summable functions on a set $A \in \mathfrak{M}$ are denoted by $\mathfrak{m}(A)$, $\bar{\mathfrak{L}}(A)$, and $\mathfrak{L}(A)$, respectively.

(14.2) Proposition. Let $A, B \in \mathfrak{M}$, $B \subset A$, let f be a function on A, and let g be the restriction of f to B. Then we have

(a) $f \in \mathfrak{m}(A) \Rightarrow g \in \mathfrak{m}(B)$
(b) $f \in \bar{\mathfrak{L}}(A) \Rightarrow g \in \bar{\mathfrak{L}}(B)$
(c) $f \in \mathfrak{L}(A) \Rightarrow g \in \mathfrak{L}(B)$

Proof. If $f\hat{}$ and $g\hat{}$ are normal extensions of f and g, respectively, then $g\hat{} = f\hat{}\chi_B$. The proposition follows from (7.18), (12.19), (12.14), (10.10), and (10.9). ///

(14.3) Definition. Let $A \in \mathfrak{M}$, $f \in \bar{\mathfrak{L}}(A)$, and let $f\hat{}$ be the normal extension of f. We set

$$\int_A f\,d\mu = \int f\hat{}\,d\mu$$

and we shall call this number the abstract Lebesgue integral of f over A.

If $f \in \bar{\mathfrak{L}}(X)$, then for the sake of consistency we shall write $\int_X f\,d\mu$ instead of $\int f\,d\mu$. With this convention for each $A \in \mathfrak{M}$, $\bar{\mathfrak{L}}(A)$ consists of all functions $f \in \mathfrak{m}(A)$ for which $\int_A f\,d\mu$ exists, and $\mathfrak{L}(A)$ consists of all functions $f \in \bar{\mathfrak{L}}(A)$ for which $\int_A f\,d\mu$ is finite.

Let $A, B \in \mathfrak{M}$, $B \subset A$, and $f \in \bar{\mathfrak{L}}(A)$. If g is the restriction of f to B, then by (14.2)(b), $\int_B g\,d\mu$ exists. Since there is no danger of confusion, we shall usually write $\int_B f\,d\mu$ instead of $\int_B g\,d\mu$.

(14.4) Proposition. Let $A_1, \ldots, A_p$ be disjoint, measurable sets, $A = \bigcup_{n=1}^p A_n$, and let f be a function on A. Then

$$\int_A f\,d\mu = \sum_{n=1}^{p} \int_{A_n} f\,d\mu$$

whenever either side has meaning.

Proof. Let $\Sigma_{n=1}^{p} \int_{A_n} f\,d\mu$ have meaning. If $f^\wedge$ is the normal extension of f, then

$$f^\wedge = f^\wedge \chi_A = \sum_{n=1}^{p} f^\wedge \chi_{A_n}$$

According to our assumption $f^\wedge \chi_{A_n} \in \mathfrak{m}(X)$, $n = 1, \ldots, p$, and so by (7.13), $f^\wedge \in \mathfrak{m}(X)$. Using (12.19) and (6.15) we obtain

$$\int_X f^{\wedge +}\,d\mu = \sum_{n=1}^{p} \int f^{\wedge +}\chi_{A_n}\,d\mu$$

and

$$\int_X f^{\wedge -} d\mu = \sum_{n=1}^{p} \int_X f^{\wedge -}\chi_{A_n}\,d\mu$$

By our assumption it follows that

$$\sum_{n=1}^{p} \int_{A_n} f\,d\mu = \sum_{n=1}^{p} \int_X f^\wedge \chi_{A_n}\,d\mu = \sum_{n=1}^{p} \int_X f^{\wedge +}\chi_{A_n}\,d\mu - \sum_{n=1}^{p} \int_X f^{\wedge -}\chi_{A_n}\,d\mu$$

$$= \int_X f^{\wedge +}\,d\mu - \int_X f^{\wedge -}\,d\mu = \int_X f^\wedge\,d\mu = \int_A f\,d\mu$$

Let $\int_A f\,d\mu$ have meaning. Applying (14.2)(b) and (c) to f^+ and f^- we obtain that $\Sigma_{n=1}^{p} \int_{A_n} f\,d\mu$ has meaning. The proposition follows from the first part of the proof. ///

(14.5) Proposition. Let $\{A_n\}$ be a sequence of disjoint measurable sets, $A = \bigcup_{n=1}^{\infty} A_n$, and let f be a function on A. Then

$$\int_A f\,d\mu = \sum_{n=1}^{\infty} \int_{A_n} f\,d\mu$$

whenever the left side has meaning.

Proof. Let $\int_A f\,d\mu$ exist. Then by (14.2)(b) the integrals $\int_{A_n} f\,d\mu$, $n = 1, 2, \ldots$, also exist. If $f^\wedge$ is the normal extension of f, then

$$f^\wedge = f^\wedge\chi_A = \sum_{n=1}^{\infty} f^\wedge\chi_{A_n}$$

It follows from (12.19) and (6.18) that

$$\int_X f^{\wedge+}\,d\mu = \sum_{n=1}^{\infty}\int_X f^{\wedge+}\chi_{A_n}\,d\mu$$

and

$$\int_X f^{\wedge-}\,d\mu = \sum_{n=1}^{\infty}\int_X f^{\wedge-}\chi_{A_n}\,d\mu$$

Therefore,

$$\int_A f\,d\mu = \int_X f^\wedge\,d\mu = \int_X f^{\wedge+}\,d\mu - \int_X f^{\wedge-}\,d\mu$$

$$= \sum_{n=1}^{\infty}\int_X f^{\wedge+}\chi_{A_n}\,d\mu - \sum_{n=1}^{\infty}\int_X f^{\wedge-}\chi_{A_n}\,d\mu$$

$$= \sum_{n=1}^{\infty}\int_X f^\wedge\chi_{A_n}\,d\mu = \sum_{n=1}^{\infty}\int_{A_n} f\,d\mu \quad ///$$

We note that $\Sigma_{n=1}^{\infty}\int_{A_n} f\,d\mu$ may have meaning, yet the integral $\int_A f\,d\mu$ may not exist [see (14-9)].

(14.6) Proposition. Let $\{A_n\} \subset \mathfrak{M}$, $A_n \nearrow A$, and let $f \in \bar{\mathfrak{L}}(A)$. Then

$$\int_A f\,d\mu = \lim \int_{A_n} f\,d\mu$$

Proof. By (14.2)(b), $\int_{A_n} f\,d\mu$ exists for $n = 1, 2, \ldots$. If $f^\wedge$ is the normal extension of f, then

$$f^{\wedge +}\chi_{A_n} \nearrow f^{\wedge +} \quad \text{and} \quad f^{\wedge -}\chi_{A_n} \nearrow f^{\wedge -}$$

By (12.19) and (6.17),

$$\int_A f\,d\mu = \int_X f^{\wedge}\,d\mu = \int_X f^{\wedge +}\,d\mu - \int_X f^{\wedge -}\,d\mu$$

$$= \lim \int_X f^{\wedge +}\chi_{A_n}\,d\mu - \lim \int_X f^{\wedge -}\chi_{A_n}\,d\mu$$

$$= \lim \int_X f^{\wedge}\chi_{A_n}\,d\mu = \lim \int_{A_n} f\,d\mu \quad ///$$

(14.7) Proposition. Let $\{A_n\} \subset \mathfrak{M}$, $A_n \searrow A$, and let $f \in \mathcal{L}(A_1)$. Then

$$\int_A f\,d\mu = \lim \int_{A_n} f\,d\mu$$

Proof. By (14.2)(c), $f \in \mathcal{L}(A)$ and $f \in \mathcal{L}(A_n)$ for $n = 1, 2, \ldots$. If $f^{\wedge}$ is the normal extension of f, then $|f^{\wedge}\chi_{A_n}| \le |f^{\wedge}|$, $n = 1, 2, \ldots$, and $\lim f^{\wedge}\chi_{A_n} = f^{\wedge}\chi_A$. Thus by (12.19) and (5.17),

$$\int_A f\,d\mu = \int_X f^{\wedge}\chi_A\,d\mu = \lim \int_X f^{\wedge}\chi_{A_n}\,d\mu = \lim \int_{A_n} f\,d\mu \quad ///$$

In the previous proposition the assumption $f \in \mathcal{L}(A_1)$ cannot be replaced by a weaker assumption $f \in \bar{\mathcal{L}}(A)$ [see (14-10)].

(14.8) Proposition. Let $A \in \mathfrak{M}$ and $f \in \mathfrak{m}(A)$. If $\mu(A) = 0$, then $f \in \mathcal{L}(A)$ and $\int_A f\,d\mu = 0$.

Proof. If $f^{\wedge}$ is the normal extension of f, then by (12.14), $f^{\wedge} \doteq 0$ [see (11.1)]. The proposition follows from (12.19), (11.2), and (4.2). ///

(14.9) Corollary. Let $A \in \mathfrak{M}$, $\mu(A) = 0$, and let f be an arbitrary function on A. If the measure μ is complete, then $f \in \mathcal{L}(A)$ and $\int_A f\,d\mu = 0$.

Proof. From the completeness of μ it easily follows that $f \in \mathfrak{m}(A)$. ///

For a noncomplete measure μ the previous corollary is generally false. It may happen that the integral $\int_A f\, d\mu$ does not exist because f does not belong to $\mathfrak{m}(A)$.

(14.10) Proposition. Let $A \in \mathfrak{M}$, $f \in \bar{\mathfrak{L}}(A)$, and let $\int_B f\, d\mu = 0$ for each measurable set $B \subset A$. Then $\mu(\{x \in A : f(x) \neq 0\}) = 0$.

Proof. Let $\mu(\{x \in A : f(x) \neq 0\}) > 0$. Without loss of generality we may assume that the set

$$B = \{x \in A : f(x) > 0\}$$

has a positive measure. Letting

$$B_n = \{x \in A : f(x) > \frac{1}{n}\}$$

$n = 1, 2, \ldots$, we have $B_n \nearrow B$. Thus by (8.5) there is an integer $N \geq 1$ such that $\mu(B_N) > 0$. Therefore,

$$\int_{B_N} f\, d\mu \geq \frac{1}{N} \int_X \chi_{B_N}\, d\mu = \frac{1}{N}\, \mu(B_N) > 0$$

a contradiction. ///

Exercises

In exercises (14-1) through (14-5) we shall assume that $(X, \mathfrak{M}, \mu)$ is an arbitrary measure space.

Let $A \in \mathfrak{M}$, and let P be a property of $x \in A$. We say that P holds μ-almost everywhere (abbreviated, μ-a.e.) in A, or that P(x) holds for μ-almost all $x \in A$, whenever the set

$$N = A - \{x \in A : P(x)\}$$

belongs to $\mathfrak{M}$ and $\mu(N) = 0$.

(14-1)+ Let $f \in \mathfrak{L}(X)$ and $\varepsilon > 0$. Show that there is a $\delta > 0$ such that $|\int_A f\,d\mu| < \varepsilon$ for each $A \in \mathfrak{M}$ for which $\mu(A) < \delta$. Hint: Suppose that for $n = 1, 2, \ldots$ there is an $A_n \in \mathfrak{M}$ such that $\mu(A_n) < 2^{-n}$ and $\int_{A_n} |f|\,d\mu \geq \varepsilon$. Set $A = \limsup A_n$ [see (7-2)] and obtain a contradiction by showing that $\mu(A) = 0$ and $\int_A |f|\,d\mu \geq \varepsilon$.

(14-2)+ Let $A \in \mathfrak{M}$, $\mu(A) < +\infty$, and let $\{f_n\} \subset \mathfrak{L}(A)$ be a sequence of finite functions which converges uniformly μ-a.e. in A to a finite function f on A.

(i) Show that $f \in \mathfrak{L}(A)$ and $\int_A f\,d\mu = \lim \int_A f_n\,d\mu$. Hint: Observe that finite constant functions on A belong to $\mathfrak{L}(A)$ and use (12.19) and (5.17).

(ii) Show by example that assertion (i) is generally false if $\mu(A) = +\infty$. Hint: Use (5-10)(iv).

(14-3)+ Let Y be a first countable topological space, $A \in \mathfrak{M}$, and let f be a function on $A \times Y$ such that

(a) $f(-,y)$ is measurable on A for each $y \in Y$

(b) There is a $g \in \mathfrak{L}(A)$ such that $|f(-,y)| \leq g$ μ-a.e. in A for each $y \in Y$

(c) $f(x,-)$ is continuous on Y for μ-almost all $x \in A$.

Show that

(i) $f(-,y)$ is summable on A for each $y \in Y$.

(ii) If $F(y) = \int_A f(-,y)\,d\mu$ for each $y \in Y$, then $F \in C(Y)$. Hint: Use the first countability of Y to apply (5.17).

(14-4)*+ Let $G \subset \mathbf{R}$ be an open interval, $A \in \mathfrak{M}$, and let f be a function on $A \times G$ such that

(a) $f(-,t)$ is measurable on A for each $t \in G$

(b) $f(-,t_o)$ is summable on A for some $t_o \in G$

(c) There is an $N \in \mathfrak{M}$ such that $\mu(N) = 0$ and $(\partial/\partial t)f(x,t)$ exists for each $t \in G$ and each $x \in A - N$

(d) There is a $g \in \mathcal{L}(A)$ such that $|(\partial/\partial t)f(x,t)| \le g(x)$ for each $t \in G$ and each $x \in A - N$.

For each $t \in G$, set $(\partial/\partial t)f(-,t)$ equal to zero on $A \cap N$ and show that

(i) $(\partial/\partial t)f(-,t)$ is summable on A for each $t \in G$. Hint: Use (7.15) and (7.11) to show that for each $t \in G$, $(\partial/\partial t)f(-,t)$ is measurable on A, and then apply (d).

(ii) $f(-,t)$ is summable on A for each $t \in G$. Hint: For $x \in A - N$ and $t \in G$,

$$|f(x,t) - f(x,t_o)| = |t - t_o| \left|\frac{\partial}{\partial t} f(x,\tau)\right|$$

(iii) If $F(t) = \int_A f(-,t)\, d\mu$ for each $t \in G$, then F is differentiable in G and

$$F'(t) = \int_A \frac{\partial}{\partial t} f(-,t)\, d\mu$$

for each $t \in G$. Hint: Observe that for $t \in G$ and small $h \neq 0$,

$$\frac{1}{h}[F(t+h) - F(t)] = \int_A \frac{1}{h}[f(-, t+h) - f(-,t)]\, d\mu$$

and apply (14-3)(ii) with respect to h.

(14-5)$^+$ Let $A \in \mathfrak{M}$, $\mathfrak{M}_A = \{B \in \mathfrak{M} : B \subset A\}$, and let μ_A be the restriction of μ to $\mathfrak{M}_A$. Show that

(i) $(A, \mathfrak{M}_A, \mu_A)$ is a measure space.

(ii) A function f on A belongs to $\mathfrak{m}(A)$, $\mathcal{L}(A)$, or $\bar{\mathcal{L}}(A)$ iff it is $\mathfrak{M}_A$-measurable, μ_A-summable, or μ_A-integrable, respectively.

(iii) If $f \in \bar{\mathcal{L}}(A)$, then $\int_A f\, d\mu = \int_A f\, d\mu_A$.

In exercises (14-6) through (14-10) we shall assume that $X = \mathbf{R}$, $\mathfrak{M} = \Lambda$ is the σ-algebra of all Lebesgue measurable subsets of $\mathbf{R}$, and $\mu = \lambda$ is the Lebesgue measure on Λ [see (10-7)].

(14-6) Let f be a continuous function on $[a,b] \times [c,d]$, where $a,b,c,d \in \mathbf{R}$, and let

$$F(y) = \int_a^b f(x,y)\, dx$$

for each $y \in [c,d]$. Apply (14-3) and (14-4) to show that

(i) F is continuous on $[c,d]$.

(ii) If $\partial f/\partial y$ is continuous on $[a,b] \times (c,d)$, then F is continuously differentiable in (c,d) and

$$F'(y) = \int_a^b \frac{\partial}{\partial y} f(x,y)\, dx$$

for each $y \in (c,d)$.

(14-7) For real numbers α, β, x, $x \neq 0$, let

$$f(x,\alpha,\beta) = e^{-\alpha x} \frac{\sin \beta x}{x}$$

and show that

(i) $f(-,\alpha,\beta)$ is measurable on $(0,+\infty)$.

(ii) If $\alpha > 0$ then $f(-,\alpha,\beta)$ is summable on $(0,+\infty)$.

(iii) If $\alpha \leq 0$ and $\beta \neq 0$, then $f(-,\alpha,\beta)$ is not integrable on $(0,+\infty)$.

Hint: Observe that

$$\int_{(0,+\infty)} f^+(-,\alpha,\beta)\, d\lambda = \int_{(0,+\infty)} f^-(-,\alpha,\beta)\, d\lambda = +\infty$$

(14-8)* Let $f(x,\alpha,\beta)$ be the function from (14-7), and let $\alpha > 0$. Carefully justifying all steps, evaluate the integral

$$K(\alpha,\beta) = \int_{(0,+\infty)} f(-,\alpha,\beta)\, d\lambda$$

Hint: Apply (14-4) to evaluate $(\partial/\partial\alpha)K(\alpha,\beta)$.

(14-9) If $A = [0,+\infty)$ and $A_n = [n-1, n)$, $n = 1, 2, \ldots,$ then A_n are disjoint and $\bigcup_{n=1}^{\infty} A_n = A$. For $x \in A$, let $f(x) = \sin 2\pi x$ and show that $\Sigma_{n=1}^{\infty} \int_{A_n} f\, d\lambda = 0$ and that the integral $\int_A f\, d\lambda$ does not exist.

(14-10) If $A_n = [n, +\infty)$, $n = 1, 2, \ldots$, then $A_n \searrow \emptyset$. For $x \in A_1$, let $f(x) = 1$ and show that $\int_{A_n} f\, d\lambda = +\infty$, $n = 1, 2, \ldots$, while $\int_{\emptyset} f\, d\lambda = 0$.

In exercise (14-11) and (14-12) we shall assume that $(X, \mathfrak{M}, \mu)$ is a measure space.

(14-11) Let f be a nonnegative measurable function on X, and for each $A \in \mathfrak{M}$, let $\nu(A) = \int_A f\, d\mu$. Show that

(i) ν is a measure on $\mathfrak{M}$.

(ii) If $f < +\infty$ and μ is σ-finite, then ν is σ-finite.

(iii) If $f > 0$ and ν is σ-finite, then μ is σ-finite.

(14-12)$^+$ Let A be an arbitrary subset of X. Prove the following:

(i) The family $\mathfrak{M}_1 = \{A \cap B : B \in \mathfrak{M}\}$ is a σ-algebra in A.

(ii) Let μ_e be the outer measure induced by μ [see (12-9)(ii)], and let μ_1 be the restriction of μ_e to $\mathfrak{M}_1$. Then $(A, \mathfrak{M}_1, \mu_1)$ is a measure space. Hint: Use (12-6)(iii).

(iii) If $A \in \mathfrak{M}$, then $(A, \mathfrak{M}_1, \mu_1)$ coincides with the measure space $(A, \mathfrak{M}_A, \mu_A)$ from (14-5)(i).

15. THE ABSTRACT FUBINI THEOREM

The Fubini theorem is a substantial generalization of the well-known formula from calculus which expresses multiple integrals as iterated integrals.

Throughout this chapter we shall assume that for $i = 1, 2$, X_i is an arbitrary set, $\mathfrak{F}_i$ is a fundamental system on X_i, and I_i is an integral on $\mathfrak{F}_i$. We shall also assume that the integrals I_1 and I_2 are either both Daniell or both Bourbaki integrals. By $\mathcal{L}_i$, $i = 1, 2$, we shall denote the family of all I_i-summable functions on X_i.

Let $X = X_1 \times X_2$, and let f be a function on X. For $x \in X_1$ and $y \in X_2$, we shall define the functions f_x on X_2 and f^y on X_1 by the following rules:

$$f_x(v) = f(x, v)$$

for each $v \in X_2$,

$$f^y(u) = f(u, y)$$

for each $u \in X_1$.

Let f be a function on X such that $f_x \in \mathcal{L}_2$ for each $x \in X_1$ and $f^y \in \mathcal{L}_1$ for each $y \in X_2$. Then setting

$$(I_2 f)(x) = I_2(f_x)$$

for each $x \in X_1$

$$(I_1 f)(y) = I_1(f^y)$$

for each $y \in X_2$, we have defined functions $I_2 f$ on X_1 and $I_1 f$ on X_2.

For an arbitrary function f on X we shall define functions $\underline{I}_2 f$ and $\overline{I}_2 f$ on X_1 and $\underline{I}_1 f$ and $\overline{I}_1 f$ on X_2 as follows:

$$(\underline{I}_2 f)(x) = \underline{I}_2(f_x) \qquad (\bar{I}_2 f)(x) = \bar{I}_2(f_x)$$

for each $x \in X_1$, and

$$(\underline{I}_1 f)(y) = \underline{I}_1(f^y) \qquad (\bar{I}_1 f)(y) = \bar{I}_1(f^y)$$

for each $y \in X_2$.

(15.1) Definition. A fundamental system $\mathfrak{F}$ on X is called a product system with respect to $(\mathfrak{F}_1, I_1)$ and $(\mathfrak{F}_2, I_2)$ whenever for each $f \in \mathfrak{F}$ the following conditions are satisfied:

(i) $f_x \in \mathfrak{F}_2$ for each $x \in X_1$ and $f^y \in \mathfrak{F}_1$ for each $y \in X_2$

(ii) $I_2 f \in \mathfrak{F}_1$ and $I_1 f \in \mathfrak{F}_2$

(iii) $I_1(I_2 f) = I_2(I_1 f)$.

The collection of all product systems with respect to $(\mathfrak{F}_1, I_1)$ and $(\mathfrak{F}_2, I_2)$ is denoted by $(\mathfrak{F}_1, I_1) \times (\mathfrak{F}_2, I_2)$. This collection is nonempty, for the fundamental system on X consisting only of the function identically equal to zero is, clearly, a product system with respect to any $(\mathfrak{F}_1, I_1)$ and $(\mathfrak{F}_2, I_2)$. On the other hand, the existence of a nontrivial system in $(\mathfrak{F}_1, I_1) \times (\mathfrak{F}_2, I_2)$ is, in general, not easy to decide.

Throughout this chapter we shall assume that $\mathfrak{F}$ is a product system with respect to $(\mathfrak{F}_1, I_1)$ and $(\mathfrak{F}_2, I_2)$, and we shall define a functional $I : \mathfrak{F} \to \mathbf{R}$ by the rule

$$If = I_1(I_2 f) = I_2(I_1 f)$$

for each $f \in \mathfrak{F}$.

(15.2) Proposition. The functional I is an integral on $\mathfrak{F}$. This integral is a Daniell or a Bourbaki integral according to whether I_1 and I_2 were Daniell or Bourbaki integrals, respectively.

Proof. Clearly, I is a nonnegative linear functional on $\mathfrak{F}$.

(a) Suppose that I_1 and I_2 are Bourbaki integrals, and let $\Phi \subset \mathfrak{F}$ and $\Phi \searrow 0$. Then for each $x \in X_1$, $\Phi_x = \{\varphi_x : \varphi \in \Phi\} \subset \mathfrak{F}_2$ and $\Phi_x \searrow 0$. Thus if

$$\Psi = \{I_2\varphi : \varphi \in \Phi\}$$

we have $\Psi \subset \mathfrak{F}_1$ and $\Psi \searrow 0$. Hence,

$$\inf\{I\varphi : \varphi \in \Phi\} = \inf\{I_1\psi : \psi \in \Psi\} = 0$$

and I is a Bourbaki integral on $\mathfrak{F}$.

(b) The case when I_1 and I_2 are Daniell integrals is quite analogous to (a); we shall leave it to the reader. ///

(15.3) Definition. The integral I on $\mathfrak{F}$ is called the product of the integrals I_1 on $\mathfrak{F}_1$ and I_2 on $\mathfrak{F}_2$.

The usage of the term "product" will be justified later by Proposition (15.8) and Corollary (15.9).

Let $\mathfrak{L}$ denote the family of all I-summable functions on X.

(15.4) Lemma. If $f \in \mathfrak{F}_+$, then

(α) $f_x \in \mathfrak{F}_{2+}$ for each $x \in X_1$ and $f^y \in \mathfrak{F}_{1+}$ for each $y \in X_2$

(β) $I_2 f \in \mathfrak{F}_{1+}$ and $I_1 f \in \mathfrak{F}_{2+}$

(γ) $If = I_1(I_2 f) = I_2(I_1 f)$.

Proof. We shall assume that I_1 and I_2 are Bourbaki integrals and that $f \in \mathfrak{F}_+^{\#}$. The case when I_1 and I_2 are Daniell integrals and $f \in \mathfrak{F}_+^{*}$ is analogous; we shall leave it to the reader. Choose $\Phi \subset \mathfrak{F}$ so that $\Phi \nearrow f$. Letting

$$\Phi_x = \{\varphi_x : \varphi \in \Phi\} \qquad \text{and} \qquad \Phi^y = \{\varphi^y : \varphi \in \Phi\}$$

where $x \in X_1$ and $y \in X_2$, we obtain that $\Phi_x \nearrow f_x$, $\Phi^y \nearrow f^y$ and $I_2\Phi_x \nearrow I_2 f_x$, $I_1\Phi^y \nearrow I_1 f^y$. From this, assertions (α) and (β) follow. Moreover,

$$\begin{aligned} If &= \sup\{I\varphi : \varphi \in \Phi\} = \sup\{I_1(I_2\varphi) : \varphi \in \Phi\} \\ &= I_1(\sup\{I_2\varphi : \varphi \in \Phi\}) = I_1(I_2 f) \end{aligned}$$

and similarly $If = I_2(I_1 f)$. ///

(15.5) Lemma. If f is a function on X, then

$$\overline{I}f \geq \max\{\overline{I}_1(\overline{I}_2 f),\ \overline{I}_2(\overline{I}_1 f)\}$$

Proof. If $g \in \mathfrak{F}_+$ and $g \geq f$, then by (15.4),

$$Ig = I_1(I_2 g) \geq \overline{I}_2(\overline{I}_2 f) \qquad \text{and} \qquad Ig = I_2(I_1 g) \geq \overline{I}_2(\overline{I}_1 f)$$

The lemma follows. ///

(15.6) Lemma. Let $f \in \mathcal{L}$. Then $\underline{I}_2 f$ and $\overline{I}_2 f$ belong to $\mathcal{L}_1$, $\underline{I}_1 f$ and $\overline{I}_1 f$ belong to $\mathcal{L}_2$, and

$$If = I_1(\underline{I}_2 f) = I_1(\overline{I}_2 f) = I_2(\underline{I}_1 f) = I_2(\overline{I}_1 f)$$

Proof. From (15.5) and its dual we obtain

$$\overline{I}f \geq \overline{I}_1(\overline{I}_2 f) \geq \overline{I}_1(\underline{I}_2 f) \geq \underline{I}_1(\underline{I}_2 f) \geq \underline{I}f$$

$$\overline{I}f \geq \overline{I}_1(\overline{I}_2 f) \geq \underline{I}_1(\overline{I}_2 f) \geq \underline{I}_1(\underline{I}_2 f) \geq \underline{I}f$$

By our assumption

$$If = \underline{I}f = \overline{I}f \neq +\infty$$

Therefore, $\underline{I}_2 f$ and $\overline{I}_2 f$ belong to $\mathcal{L}_1$, and

$$If = I_1(\underline{I}_2 f) = I_1(\overline{I}_2 f)$$

The rest of the proof is similar. ///

If the system $\mathcal{L}_i$, $i = 1, 2$, is Stonian [see (10.1)], we shall denote by $(X_i, \mathfrak{I}_i, \iota_i)$ the measure space induced by the integral I_i [see (10.4)]. Recall that by (10.2), the system $\mathcal{L}_i$ is Stonian whenever the fundamental system $\mathfrak{F}_i$ is.

(15.7) Theorem (Fubini). Let $\mathcal{L}_1$ and $\mathcal{L}_2$ be Stonian systems, and let $f \in \mathcal{L}$. Then $f_x \in \mathcal{L}_2$ for ι_1-almost all $x \in X_1$ and $f^y \in \mathcal{L}_1$ for ι_2-almost all $y \in X_2$. Moreover if F and G are functions on X_1 and X_2, respectively, such that

$$F(x) = I_2(f_x) \qquad \text{and} \qquad G(y) = I_1(f^y)$$

whenever the integrals on the right side exist, then $F \in \mathcal{L}_1$, $G \in \mathcal{L}_2$, and

$$If = I_1 F = I_2 G$$

Proof. For $y \in X_2$, set

$$h(y) = \bar{I}_1(f^y) - \underline{I}_1(f^y)$$

if $\bar{I}_1(f^y) - \underline{I}_1(f^y)$ has meaning, and $h(y) = 0$ otherwise. By (15.6), (5.14), and (5.15), $h \in \mathcal{L}_2$ and

$$I_2 h = I_2(\bar{I}_1 f) - I_2(\underline{I}_1 f) = 0$$

Since $h \geq 0$, it follows from (11.7) that ι_2-almost everywhere $h = 0$. Thus by (15.6) and (11.5),

$$\underline{I}_1(f^y) = \bar{I}_1(f^y) \neq \pm\infty$$

for ι_2-almost all $y \in X_2$. It follows that $f^y \in \mathcal{L}_1$ and $I_1(f^y) = G(y)$ for ι_2-almost all $y \in X_2$. In particular, ι_2-almost everywhere $G = \bar{I}_1 f$. By (15.6) and (11.2), $G \in \mathcal{L}_2$ and $If = I_2 G$. The rest of the theorem follows by symmetry. ///

In view of (11.10), the assumption that $\mathcal{L}_1$ and $\mathcal{L}_2$ are Stonian systems is merely a matter of convenience; the previous theorem remains correct without this assumption provided the expressions "ι_i-almost all," $i = 1, 2$, are properly defined.

If g and h are functions on X_1 and X_2 , respectively, we define the function $g \otimes h$ on X by setting

$$(g \otimes h)(x,y) = g(x)h(y)$$

for each $(x,y) \in X$. We shall denote by $\mathfrak{F}_1 \otimes \mathfrak{F}_2$ the vector space of functions on X generated by the family

$$\{g \otimes h : g \in \mathfrak{F}_1 , h \in \mathfrak{F}_2\}$$

The reader familiar with algebra will recognize that $\mathfrak{F}_1 \otimes \mathfrak{F}_2$ is isomorphic to the tensor product of the vector spaces $\mathfrak{F}_1$ and $\mathfrak{F}_2$.

(15.8) Proposition. Suppose that $\mathfrak{F}_1 \otimes \mathfrak{F}_2 \subset \mathfrak{F}$, and let g and h be nonnegative functions on X_1 and X_2, respectively. If $f = g \otimes h$, then

$$\bar{I}f = (\bar{I}_1 g)(\bar{I}_2 h) \qquad \text{and} \qquad \underline{I}f = (\underline{I}_1 g)(\underline{I}_2 h)$$

unless the product on the right side is of the form $0 \cdot (+\infty)$ or $(+\infty) \cdot 0$.

Proof. (a) Since $f_x = g(x)h$, it follows from (5.3) that

$$\bar{I}_2 f = (\bar{I}_2 h) g \qquad \text{and} \qquad \bar{I}_1(\bar{I}_2 f) = (\bar{I}_1 g)(\bar{I}_2 h)$$

Thus by (15.5),

$$\bar{I}f \geq \bar{I}_1(\bar{I}_2 f) = (\bar{I}_1 g)(\bar{I}_2 h)$$

Hence it suffices to show that

$$\bar{I}f \leq (\bar{I}_1 g)(\bar{I}_2 h)$$

provided $\bar{I}_1 g < +\infty$ and $\bar{I}_2 h < +\infty$. We shall assume that I_1 and I_2 are Bourbaki integrals. The case when I_1 and I_2 are Daniell integrals is again completely analogous, and we shall leave it to the reader. Thus suppose that $\bar{I}_1 g < +\infty$ and $\bar{I}_2 h < +\infty$, and choose $g_o \in \mathfrak{F}^{\#}_{1+}$, $h_o \in \mathfrak{F}^{\#}_{2+}$ so that $g_o \geq g$ and $h_o \geq h$. Let

$$\Phi = \{\varphi \in \mathfrak{F}_1 : 0 \leq \varphi \leq g_o\}$$

$$\Psi = \{\psi \in \mathfrak{F}_2 : 0 \leq \psi \leq h_o\}$$

$$\Gamma = \{\varphi \otimes \psi : \varphi \in \Phi,\ \psi \in \Psi\}$$

Because g_o and h_o are nonnegative, by (2.16), $\Phi \nearrow g_o$ and $\Psi \nearrow h_o$. Hence, $\Gamma \nearrow g_o \otimes h_o$, and since $\mathfrak{F}_1 \otimes \mathfrak{F}_2 \subset \mathfrak{F}$, we have $g_o \otimes h_o \in \mathfrak{F}^{\#}_{+}$. Clearly, $g_o \otimes h_o \geq f$, and so

$$\begin{aligned}\bar{I}f \leq I(g_o \otimes h_o) &= \sup\{I(\varphi \otimes \psi) : \varphi \in \Phi,\ \psi \in \Psi\}\\ &= \sup\{(I_1\varphi)(I_2\psi) : \varphi \in \Phi,\ \psi \in \Psi\}\\ &= \sup\{I_1\varphi : \varphi \in \Phi\}\sup\{I_2\psi : \psi \in \Psi\} = (I_1 g_o)(I_2 h_o)\end{aligned}$$

It follows from the arbitrariness of g_o and h_o that

$$\bar{I}f \leq (\bar{I}_1 g)(\bar{I}_2 h)$$

(b) The proof for the lower integrals is quite similar to (a), and we shall leave it to the reader as an exercise. ///

The assumption about the products

$$(\bar{I}_1 g)(\bar{I}_2 h) \qquad \text{and} \qquad (\underline{I}_1 g)(\underline{I}_2 h)$$

being of the form different from $0 \cdot (+\infty)$ and $(+\infty) \cdot 0$ is essential in the previous proposition [see (15-3), (16-4), and (17-9)(i)].

(15.9) Corollary. Suppose that $\mathfrak{F}_1 \otimes \mathfrak{F}_2 \subset \mathfrak{F}$, and let $g \in \mathcal{L}_1$ and $h \in \mathcal{L}_2$. If $f = g \otimes h$, then $f \in \mathcal{L}$ and

$$If = (I_1 g)(I_2 h)$$

Proof. Since

$$f = g^+ \otimes h^+ - g^+ \otimes h^- - g^- \otimes h^+ + g^- \otimes h^-$$

the corollary follows from (15.8), (5.14), and (5.15). ///

Proposition (15.8) and Corollary (15.9) motivate the following definition.

(15.10) Definition. A system $\mathfrak{F} \in (\mathfrak{F}_1, I_1) \times (\mathfrak{F}_2, I_2)$ is called ample if $\mathfrak{F}_1 \otimes \mathfrak{F}_2 \subset \mathfrak{F}$.

From Corollary (15.9) it is obvious that the important product systems are those which are ample. Unfortunately, it is not clear, in general, whether an ample product system with respect to $(\mathfrak{F}_1, I_1)$ and $(\mathfrak{F}_2, I_2)$ exists. The vector space $\mathfrak{F}_1 \otimes \mathfrak{F}_2$, which is an obvious candidate for an ample system in $(\mathfrak{F}_1, I_1) \times (\mathfrak{F}_2, I_2)$, may not be a fundamental system on X; for $\mathfrak{F}_1 \otimes \mathfrak{F}_2$ is generally not a vector lattice [see (15-8)(ii)]. We have, however, the following useful proposition.

(15.11) Proposition. If $|f| \in \mathfrak{F}_1 \otimes \mathfrak{F}_2$ whenever $f \in \mathfrak{F}_1 \otimes \mathfrak{F}_2$, then $\mathfrak{F}_1 \otimes \mathfrak{F}_2$ is an ample product system with respect to $(\mathfrak{F}_1, I_1)$ and $(\mathfrak{F}_2, I_2)$.

The proof is straightforward, and we leave it to the reader.

The possible lack of ample product systems with respect to $(\mathfrak{F}_1, I_1)$ and $(\mathfrak{F}_2, I_2)$ appears to be an unpleasant feature. Fortunately, there is a way out whenever the systems $\mathcal{L}_1$ and $\mathcal{L}_2$ are Stonian [see (10.1)]. If $\mathcal{L}_i$, i = 1, 2, is a Stonian system, let $(X_i, \mathfrak{J}_i, \iota_i)$ be the measure space induced by the integral I_i [see (10.4)]. By $\mathfrak{F}_i'$ and I_i' we shall denote, respectively, the fundamental system and the Daniell integral induced by the measure ι_i [see Chapter 12, in particular, (12.10)]. According to (12.18), the integrals I_i and I_i' coïncide when extended to the family of all integrable functions. Moreover, in the next chapter [see (16.2)] we shall show that $\mathfrak{F}_1' \otimes \mathfrak{F}_2'$ is a Stonian ample product system with respect to $(\mathfrak{F}_1', I_1')$ and $(\mathfrak{F}_2', I_2')$.

Exercises

(15-1) For i = 1, 2, let X_i be a discrete space, $\mathfrak{F}_i = C_o(X_i)$, and let

$$I_i f = \sum_{x \in X_i} f(x)$$

for each $f \in \mathfrak{F}_i$ [see (4-4)]. Set $X = X_1 \times X_2$, and show that

(i) $C_o(X) = \mathfrak{F}_1 \otimes \mathfrak{F}_2$, and so $\mathfrak{F} = C_o(X)$ is an ample product system with respect to $(\mathfrak{F}_1, I_1)$ and $(\mathfrak{F}_2, I_2)$.

(ii) If I on $\mathfrak{F}$ is the product of the integrals I_1 and I_2, then

$$If = \sum_{x \in X} f(x)$$

for each $f \in \mathfrak{F}$.

(15-2)+ Let a,b,c,d be real numbers, a < b, c < d, $X_1 = [a,b]$, $X_2 = [c,d]$, and let $X = X_1 \times X_2$. Furthermore, let $\mathfrak{F}_i = C(X_i)$, and let I_i be the Riemann integral on $\mathfrak{F}_i$, i = 1, 2. Show that $\mathfrak{F} = C(X)$ is an ample product system with respect to $(\mathfrak{F}_1, I_1)$ and $(\mathfrak{F}_2, I_2)$. Hint: Use the formula

$$\int_c^d \int_a^b f(x,y)\,dx\,dy = \int_a^b \int_c^d f(x,y)\,dy\,dx$$

for Riemann integrals, which you know from the calculus [see Ref. 1, Theorem 10-20(v), p. 261].

(15-3) Let X_1 be an uncountable discrete space, $\mathfrak{F}_1 = C_o(X_1)$, and let

$$I_1 f = \sum_{x \in X_1} f(x)$$

for each $f \in \mathfrak{F}_1$ [see (4-4)]. Let $X_2 = W$ [see Chapter 1, Section B] and $\mathfrak{F}_2 = C(X_2)$. If $f \in \mathfrak{F}_2$, then there is a $c_f \in R$ such that the set $\{x \in X_2 : f(x) \neq c_f\}$ is countable [see (1-15)(ii)]. Setting $I_2 f = c_f$ for each $f \in \mathfrak{F}_2$, we have defined a Daniell integral I_2 on $\mathfrak{F}_2$ [see (4-3)]. Show that

(i) $\mathfrak{F} = \mathfrak{F}_1 \otimes \mathfrak{F}_2$ is a product system with respect to $(\mathfrak{F}_1, I_1)$ and $(\mathfrak{F}_2, I_2)$.

(ii) Let I on $\mathfrak{F}$ be the product of the integrals I_1 and I_2. If $g = \chi_{X_1}$ and $h = \chi_A$, where A is a countable subset of X_2, then g, h, and $g \otimes h$ are, respectively, I_1, I_2, and I-integrable, and we have $I_1 g = +\infty$, $I_2 h = 0$, and $I(g \otimes h) = +\infty$. Hint: Observe that I is a Daniell but not a Bourbaki integral on $\mathfrak{F}$.

(15-4) Let X be an arbitrary set, and let $\mathfrak{G}$ be a family of real-valued functions on X. Denote by $[\mathfrak{G}]$ the vector space generated by $\mathfrak{G}$, and let $|\mathfrak{G}| = \{|f| : f \in \mathfrak{G}\}$. Define inductively vector spaces $\mathfrak{G}_n$, $n = 0, 1, \ldots$, by letting $\mathfrak{G}_0 = [\mathfrak{G} \cup |\mathfrak{G}|]$ and $\mathfrak{G}_{n+1} = [\mathfrak{G}_n \cup |\mathfrak{G}|]$. Set $\mathfrak{G}_\infty = \bigcup_{n=0}^{\infty} \mathfrak{G}_n$, and show that

(i) $\mathfrak{G}_\infty$ is a fundamental system on X containing $\mathfrak{G}$.

(ii) If $\mathfrak{F}$ is a fundamental system on X and $\mathfrak{G} \subset \mathfrak{F}$, then $\mathfrak{G}_\infty \subset \mathfrak{F}$.

Thus $\mathfrak{G}_\infty$ is the smallest fundamental system on X containing $\mathfrak{G}$. We shall say that $\mathfrak{G}_\infty$ is the fundamental system on X <u>generated</u> by $\mathfrak{G}$.

(15-5) Let X_i be an arbitrary set, and let $\mathfrak{F}_i$ be a fundamental system on X_i, $i = 1, 2$. Furthermore, let I_1 be the Dirac integral at $x_o \in X_1$ on $\mathfrak{F}_1$, and let I_2 be the Dirac integral at $y_o \in X_2$ on $\mathfrak{F}_2$ [see (4-1)]. Set $X = X_1 \times X_2$, and show that

(i) If $\mathfrak{F} = (\mathfrak{F}_1 \otimes \mathfrak{F}_2)_\infty$ is the fundamental system on X generated by $\mathfrak{F}_1 \otimes \mathfrak{F}_2$ [see (15-4)], then $\mathfrak{F}$ is an ample product system with respect to $(\mathfrak{F}_1, I_1)$ and $(\mathfrak{F}_2, I_2)$.

(ii) If I on $\mathfrak{F}$ is the product of the integrals I_1 and I_2, then I is the Dirac integral at (x_o, y_o).

(15-6) Let $\alpha_1, \ldots, \alpha_n$ be distinct elements of the interval $(0,\pi)$, and let

$$f_i(x) = |\sin(x + \alpha_i)| \qquad (i = 1, \ldots, n)$$

for each $x \in [0,\pi]$. Show that the functions $f_1, \ldots, f_n$ on $[0,\pi]$ are linearly independent. Hint: Observe that for $x \in (0,\pi)$ the derivative $f_i'(x)$ exists iff $x \neq \pi - \alpha_i$, $i = 1, \ldots, n$.

(15-7) Let $\alpha_1, \ldots, \alpha_n$ be distinct real numbers, let (a,b) be a nonempty open interval, and let

$$f_i(x) = e^{\alpha_i x} \qquad (i = 1, \ldots, n)$$

for each $x \in (a,b)$. Show that the functions $f_1, \ldots, f_n$ on (a,b) are linearly independent. Hint: Suppose $\Sigma_{i=1}^n c_i f_i = 0$ for some real numbers $c_1, \ldots, c_n$. Applying to this equation the differential operator $D_j = \Pi_{i=1, i\neq j}^n (d/dx - \alpha_i)$, $j = 1, \ldots, n$, we obtain

$$c_j f_j \prod_{\substack{i=1 \\ i\neq j}}^{n} (\alpha_j - \alpha_i) = 0$$

and it follows that $c_j = 0$.

In exercises (15-8) and (15-9) we shall assume that $X_1 = X_2 = [0,\pi]$, $X = X_1 \times X_2$, $\mathfrak{F}_1 = \mathfrak{F}_2 = C([0,\pi])$, and $\mathfrak{G} = \mathfrak{F}_1 \otimes \mathfrak{F}_2$.

(15-8)[+] Show that

(i) If $f(x,y) = |\sin(x + y)|$ for each $(x,y) \in X$, then $f \notin \mathfrak{G}$. Hint: If $f \in \mathfrak{G}$, then for each $y \in X_2$, $f^y = \Sigma_{i=1}^n h_i(y) g_i$, where $g_i \in \mathfrak{F}_1$ and $h_i \in \mathfrak{F}_2$, $i = 1, \ldots, n$. Contrary to (15-6), this implies that any family $\{f^y\}_y$ with more than n elements is linearly dependent.

(ii) $\mathfrak{G}$ is not a fundamental system on X. Hint: Observe that $\sin(x + y) = \sin x \cos y + \cos x \sin y$ and use (i).

(15-9)[+] If g is a function on X and $A \subset X$, denote by g_A the restriction of g to A. If $\mathcal{E}$ is a family of functions on X and $A \subset X$, let $\mathcal{E}_A = \{g_A : g \in$

For each $(x,y) \in X$, set $f(x,y) = e^{xy}$ and prove

(i) If an open set $G \subset X$ is nonempty, then $f_G \notin \mathfrak{a}_G$. Hint: Find nonempty open intervals $A_i \subset X_i$ so that $A_1 \times A_2 \subset G$. Use (15-7) and modify the hint in (15-8)(i).

(ii) Let $\mathfrak{a}_0 = [\mathfrak{a} \cup |\mathfrak{a}|]$ [see (15-4)]. If an open set $G \subset X$ is nonempty, then $f_G \notin (\mathfrak{a}_0)_G$. Hint: Suppose $f_G \in (\mathfrak{a}_0)_G$. Then $f_G = h_G^0 + \Sigma_{i=1}^n |h_G^i|$, where $h^0, \ldots, h^n$ belong to $\mathfrak{a}$. Find a nonempty open set $H \subset G$ on which the functions $h^1, \ldots, h^n$ do not change sign. Then $f_H \in \mathfrak{a}_H$ contrary to (i).

(iii) If $\mathfrak{a}_\infty$ is the fundamental system generated by $\mathfrak{a}$, then $f \notin \mathfrak{a}_\infty$. Hint: Show inductively that if an open set $G \subset X$ is nonempty, then $f_G \notin (\mathfrak{a}_n)_G$, $n = 0, 1, \ldots$ [see (15-4)].

(iv) $\mathfrak{a} \subsetneqq \mathfrak{a}_\infty \subsetneqq \mathcal{C}(X)$.

16. PRODUCT MEASURES

If μ_1 and μ_2 are measures in the sets X_1 and X_2, respectively, we shall use the results of Chapter 15 to construct a measure μ in the set $X_1 \times X_2$.

Throughout this chapter we shall assume that $(X_1, \mathfrak{M}_1, \mu_1)$ and $(X_2, \mathfrak{M}_2, \mu_2)$ are arbitrary measure spaces, and we set $X = X_1 \times X_2$. For $i = 1, 2$, we shall denote by $\mathfrak{F}_i$ and I_i, respectively, the fundamental system and the Daniell integral induced by the measure μ_i [see Chapter 12, in particular, (12.10)]. We let

$$\mathfrak{R} = \{A_1 \times A_2 : A_i \in \mathfrak{M}_i, \ i = 1, 2\}$$

$$\mathfrak{R}_o = \{A_1 \times A_2 \in \mathfrak{R} : \mu_i(A_i) < +\infty, \ i = 1, 2\}$$

and we shall call the sets from $\mathfrak{R}$ the <u>measurable rectangles</u>.

Let $A \subset X$, $x \in X_1$, and $y \in X_2$. The sets

$$A_x = \{v \in X_2 : (x, v) \in A\}$$

and

$$A^y = \{u \in X_1 : (u, y) \in A\}$$

are called <u>sections</u> of A at x and y, respectively (see Fig. 16.1). If f is a function on A, we shall define functions f_x on A_x and f^y on A^y by the following rules:

$$f_x(v) = f(x, v)$$

for each $v \in A_x$,

$$f^y(u) = f(u, y)$$

for each $u \in A^y$.

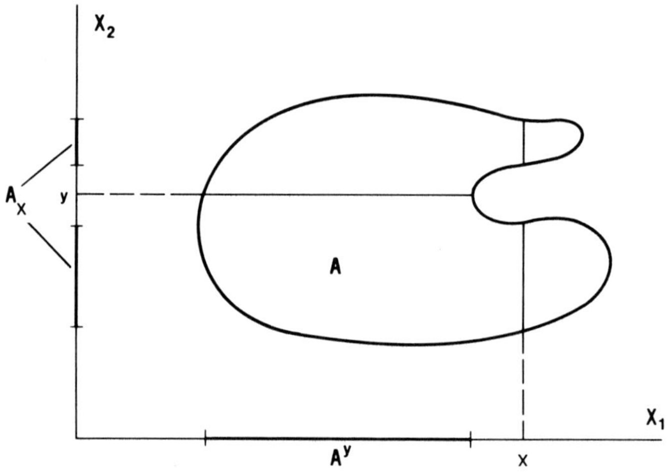

FIG. 16.1

The following observation about $\mathfrak{R}$ is closely related to Lemma (12.6). Recall that the definition of a strong refinement was given in (12.5).

(16.1) Lemma. Every finite family $\mathfrak{C} \subset \mathfrak{R}$ has a finite disjoint strong refinement $\mathfrak{D} \subset \mathfrak{R}$.

Proof. Let $\mathfrak{C} = \{A_i \times B_i\}_{i=1}^n$ be a subfamily of $\mathfrak{R}$. By (12.6), there are disjoint strong refinements $\{A_j'\}_{j=1}^p \subset \mathfrak{M}_1$ of $\{A_i\}_{i=1}^n$ and $\{B_k'\}_{k=1}^q \subset \mathfrak{M}_2$ of $\{B_i\}_{i=1}$. It suffices to let $\mathfrak{D}$ consist of those sets $A_j' \times B_k'$ (j and k are integers, $1 \leq j \leq p$, $1 \leq k \leq q$) which are contained in $\bigcup_{i=1}^n (A_i \times B_i)$. ///

(16.2) Proposition. The system $\mathfrak{F} = \mathfrak{F}_1 \otimes \mathfrak{F}_2$ is a Stonian ample product system with respect to $(\mathfrak{F}_1, I_1)$ and $(\mathfrak{F}_2, I_2)$.

Proof. Since $\chi_{A\times B} = \chi_A \otimes \chi_B$ for each $A \times B \in \mathfrak{R}$, the system $\mathfrak{F}$ consists of all functions f on X for which

$$f = \sum_{i=1}^n c_i \chi_{C_i}$$

where $c_i \in \mathbf{R}$ and $C_i \in \mathfrak{R}_o$, $i = 1, \ldots, n$. According to (16.1) we may assume that the sets $C_1, \ldots, C_n$ are disjoint. Hence,

$$|f| = \sum_{i=1}^{n} |c_i|\chi_{C_i} \qquad \text{and} \qquad 1 \wedge f = \sum_{i=1}^{n} (1 \wedge c_i)\chi_{C_i}$$

and the proposition follows from (15.11) and (10.1). ///

Let I on $\mathfrak{F} = \mathfrak{F}_1 \otimes \mathfrak{F}_2$ be the product of the integrals I_1 on $\mathfrak{F}_1$ and I_2 on $\mathfrak{F}_2$ [see (15.3)]. If $f \in \mathfrak{F}$, it is easy to calculate the integral If in terms of the measures μ_1 and μ_2. Indeed, since

$$f = \sum_{i=1}^{n} c_i(\chi_{A_i} \otimes \chi_{B_i})$$

where $c_i \in \mathbf{R}$ and $A_i \times B_i \in \mathfrak{R}_o$, $i = 1, \ldots, n$, we have

$$If = \sum_{i=1}^{n} c_i\mu_1(A_i)\mu_2(B_i)$$

The integral I is a Daniell integral, for the integrals I_1 and I_2 are. Because $\mathfrak{F}$ is a Stonian system, I induces a measure space $(X, \mathfrak{I}, \iota)$ [see (10.2) and (10.4)].

<u>(16.3)</u> <u>Definition</u>. The measure space $(X, \mathfrak{I}, \iota)$ is called the <u>product</u> of the measure spaces $(X_1, \mathfrak{M}_1, \mu_1)$ and $(X_2, \mathfrak{M}_2, \mu_2)$.

Notice that the measure ι is always complete and saturated regardless of whether the measures μ_1 and μ_2 are.

The following proposition justifies the usage of the term "product."

<u>(16.4)</u> <u>Proposition</u>. $\mathfrak{R} \subset \mathfrak{I}$, and if $A \times B \in \mathfrak{R}$, then

$$\iota(A \times B) = \mu_1(A)\mu_2(B)$$

unless the product on the right side is of the form $0 \cdot (+\infty)$ or $(+\infty) \cdot 0$.

<u>Proof</u>. Let $C \in \mathfrak{R}$, and let f be a nonnegative function from $\mathfrak{F}$. Using (16.1), we can write

$$f = \sum_{i=1}^{n} c_i\chi_{C_i}$$

where $c_1, \ldots, c_n$ are nonnegative real numbers and $C_1, \ldots, C_n$ are disjoint sets from $\mathfrak{R}_o$. Thus

$$\chi_C \wedge f = \sum_{i=1}^{n} (1 \wedge c_i)\chi_{C\cap C_i}$$

and since $C \cap C_i \in \mathfrak{R}_o$, we have $\chi_C \wedge f \in \mathfrak{F}$. By (6.7), $\chi_C \in \mathfrak{L}_+$; hence $C \in \mathfrak{I}$. Since $\chi_{A\times B} = \chi_A \otimes \chi_B$ for each $A \times B \in \mathfrak{R}$, the proposition follows from (15.8). ///

What happens in the previous proposition when the product $\mu_1(A)\mu_2(B)$ is of the form $0\cdot(+\infty)$ or $(+\infty)\cdot 0$ will be discussed in (16-4) (iv) and (v).

For i = 1, 2, let $(X_i, \mathfrak{I}_i, \iota_i)$ be the measure space induced by the measure μ_i [see (12.10)]. The next proposition shows that the products of the measure spaces $(X_i, \mathfrak{M}_i, \mu_i)$ and $(X_i, \mathfrak{I}_i, \iota_i)$ are the same.

<u>(16.5)</u> <u>Proposition</u>. The product of the measure spaces $(X_1, \mathfrak{I}_1, \iota_1)$ and $(X_2, \mathfrak{I}_2, \iota_2)$ is the measure space $(X, \mathfrak{I}, \iota)$.

<u>Proof</u>. For the purpose of this proof only denote by $\mathfrak{F}'_i$ and I'_i, respectively, the fundamental system and the Daniell integral induced by the measure ι_i, i = 1, 2. Let $\mathfrak{F}' = \mathfrak{F}'_1 \otimes \mathfrak{F}'_2$, and let I' on $\mathfrak{F}'$ be the product of integrals I'_1 on $\mathfrak{F}'_1$ and I'_2 on $\mathfrak{F}'_2$. By (12.14), $\mathfrak{F} \subset \mathfrak{F}'$ and $I'f = If$ for each $f \in \mathfrak{F}$. According to (6.21),

$$\underline{I}f \leq \underline{I}'f \leq \bar{I}'f \leq \bar{I}f$$

for every function f on X. Let $A_i \in \mathfrak{I}_i$ and $\iota_i(A_i) < +\infty$, i = 1, 2. By (12.12) and the completeness of ι, $A_1 \times A_2 \in \mathfrak{I}$ and

$$\iota(A_1 \times A_2) = \iota_1(A_1)\iota_2(A_2)$$

Hence each function $f \in \mathfrak{F}'$ is I-summable and $If = I'f$. Because I' is a Daniell integral on $\mathfrak{F}'$, this implies that each function $f \in \bar{\mathfrak{F}}'$ is I-integrable and $If = I'f$. Therefore,

$$\underline{I}'f \leq \underline{I}f \leq \bar{I}f \leq \bar{I}'f$$

for every function f on X, and the proposition follows. ///

Throughout the rest of this chapter we shall assume that the measures μ_1 and μ_2 are <u>complete</u> and we shall use the notation established in Chapter 14. Thus, e.g., if $A_i \in \mathfrak{M}_i$, $i = 1, 2$, then $\bar{\mathfrak{L}}(A_i)$ will denote the family of all functions f on A_i for which the integral $\int_{A_i} f\, d\mu_i$ exists. Similarly, if $A \in \mathfrak{J}$, then $\bar{\mathfrak{L}}(A)$ will denote the family of all functions f on A for which the integral $\int_A f\, d\iota$ exists. The meaning of the symbols $\mathfrak{L}(A_1)$, $\mathfrak{L}(A_2)$, and $\mathfrak{L}(A)$ is obvious.

(16.6) Theorem. Let $f \in \mathfrak{L}(X)$. Then $f_x \in \mathfrak{L}(X_2)$ for μ_1-almost all $x \in X_1$ and $f^y \in \mathfrak{L}(X_1)$ for μ_2-almost all $y \in X_2$. Moreover, if F and G are functions on X_1 and X_2, respectively, such that

$$F(x) = \int_{X_2} f_x\, d\mu_2 \qquad G(y) = \int_{X_1} f^y\, d\mu_1$$

whenever the integrals on the right side exist, then $F \in \mathfrak{L}(X_1)$, $G \in \mathfrak{L}(X_2)$, and

$$\int_X f\, d\iota = \int_{X_1} F\, d\mu_1 = \int_{X_2} G\, d\mu_2$$

Proof. From (12.22) we obtain that f is I-summable and $\int_X f\, d\iota = If$. According to (12.18), for $i = 1, 2$, the integral I_i and the measure space $(X_i, \mathfrak{J}_i, \iota_i)$ are mutually inducing each other. Since the measures μ_1 and μ_2 are complete, it follows from (12.19) and (12.20) that $\mathfrak{L}(X_i)$ consists of all I_i-summable functions and that $I_i h = \int_{X_i} h\, d\mu_i$ for each $h \in \mathfrak{L}(X_i)$, $i = 1, 2$. Moreover, by (12.14) and (12.16) the null ideals $\mathfrak{M}_{i,o}$ and $\mathfrak{J}_{i,o}$ coincide [see (8.7)]. Hence "μ_i-almost all" means the same as "ι_i-almost all," $i = 1, 2$. The theorem follows from (15.7). ///

The previous theorem is conceptually quite simple. To underline this simplicity we shall introduce the following notational convention.

If $A \in \mathfrak{J}$ and $f \in \bar{\mathfrak{L}}(A)$, we shall agree to understand the symbol

$$\int_{X_1}\left(\int_{A_x} f_x\, d\mu_2\right) d\mu_1$$

as follows: For μ_1-almost all $x \in X_1$, $A_x \in \mathfrak{M}_2$ and $f_x \in \bar{\mathfrak{L}}_2(A_x)$; if F is a function on X_1 such that

$$F(x) = \int_{A_x} f_x \, d\mu_2$$

whenever the integral on the right side exists, then $F \in \bar{\mathfrak{L}}(X_1)$ and

$$\int_{X_1} \left(\int_{A_x} f_x \, d\mu_2 \right) d\mu_1 = \int_{X_1} F \, d\mu_1$$

We shall adhere to this convention and its obvious symmetric counterpart throughout the remainder of this book. Theorem (16.6) will now take a very intuitive form:

If $f \in \mathfrak{L}(X)$, then

$$\int_X f \, d\mu = \int_{X_1} \left(\int_{X_2} f_x \, d\mu_2 \right) d\mu_1 = \int_{X_2} \left(\int_{X_1} f^y \, d\mu_1 \right) d\mu_2$$

<u>(16.7)</u> <u>Theorem</u> (Tonelli). Let $A \in \mathfrak{I}$ be an ι-σ-finite set, and let $f \in \bar{\mathfrak{L}}(A)$. Then

$$\int_A f \, d\iota = \int_{X_1} \left(\int_{A_x} f_x \, d\mu_2 \right) d\mu_1 = \int_{X_2} \left(\int_{A^y} f^y \, d\mu_1 \right) d\mu_2$$

<u>Proof</u>. Using the definition of a σ-finite set [see (8.13)], we can find sets $A_n \in \mathfrak{I}$ such that $\iota(A_n) < +\infty$, $A_n \subset A_{n+1}$, $n = 1, 2, \ldots$, and $A = \bigcup_{n=1}^{\infty} A_n$. Set $g(z) = f^+(z)$ if $z \in A$ and $g(z) = 0$ if $z \in X - A$, and let $g_n = g \wedge (n\chi_{A_n})$, $n = 1, 2, \ldots$. Then $g_n \in \mathfrak{L}(X)$, and so by (16.6),

$$\int_X g_n \, d\iota = \int_{X_1} \left(\int_{X_2} (g_n)_x \, d\mu_2 \right) d\mu_1$$

Clearly, g and g_x, $x \in X_1$, are the normal extensions [see (14.1)] of and f_x^+, respectively [note that $(f^+)_x = (f_x)^+$]. Also $g_n \nearrow g$ and $(g_n)_x \nearrow g_x$ for all $x \in X_1$. Therefore by (14.3) and (6.17),

$$\int_A f^+ \, d\iota = \int_X g \, d\iota = \int_{X_1} \left(\int_{X_2} g_x \, d\mu_2 \right) d\mu_1 = \int_{X_1} \left(\int_{A_x} f_x^+ \, d\mu_2 \right) d\mu_1$$

Similarly we can prove that

$$\int_A f^- \, d\iota = \int_{X_1} \left(\int_{A_x} f_x^- \, d\mu_2 \right) d\mu_1$$

and hence,

$$\int_A f \, d\iota = \int_{X_1} \left(\int_{A_x} f_x \, d\mu_2 \right) d\mu_1$$

The rest of the theorem follows by symmetry. ///

Applying the previous theorem to the function $f = 1$, we obtain the following:

(16.8) Corollary. If $A \in \mathfrak{I}$ is an ι-σ-finite set, then

$$\iota(A) = \int_{X_1} \mu_2(A_x) \, d\mu_1 = \int_{X_2} \mu_1(A^y) \, d\mu_2$$

At this point a few remarks are in place.

(i) Corollary (16.8) is written in a kind of shorthand. The equation in this corollary should be interpreted in the same way as we interpreted the equation in Theorem (16.7). We urge the reader to carry the interpretation out in detail.

(ii) Without the assumption that the measures μ_1 and μ_2 are complete, Theorems (16.6), (16.7), and Corollary (16.8) are generally false [see (16-8)].

(iii) Although Theorem (16.7) is only a mild generalization of Theorem (16.6) [that it is a generalization follows from (10.8)], it has a considerable importance for deciding whether a function f on a ι-σ-finite set $A \in \mathfrak{I}$ is ι-summable on A. Usually it is rather easy to see whether f is ι-integrable on A. For instance, if f is a nonnegative function, it suffices to show that f is $\mathfrak{I}$-measurable on A and this can frequently be done by some topological considerations. Once the ι-integrability of f on A is established, we can apply Theorem (16.7) and use the iterated integrals to evaluate, or at least estimate, the integral $\int_A f\, d\iota$. To check the summability of f on A directly is often much harder [see (16-11)].

(iv) Without assuming the ι-σ-finiteness of the set $A \in \mathfrak{J}$, Theorem (16.7) and Corollary (16.8) are generally false [see (16-4)].

(v) Warning: The existence and equality of the iterated integrals in Theorem (16.6) does not, in general, imply the existence of the product integral. The same is true of Theorem (16.7) and Corollary (16.8). Some illustrative examples will be given in (16-12) and (16-13).

Exercises

(16-1)$^+$ Let $\mathfrak{A}$ be an algebra in a set X [see (7-4)], and let $\mathfrak{M}$ be the σ-algebra in X generated by $\mathfrak{A}$ [see (7.3)]. Suppose that μ and ν are measures on $\mathfrak{M}$ such that $\mu(A) = \nu(A)$ for each $A \in \mathfrak{A}$ and show that

(i) If $\mu(X) < +\infty$, then $\mu(A) = \nu(A)$ for each $A \in \mathfrak{M}$. Hint: Using (7-6)(i) and (8-7)(iii), proceed by transfinite induction.

(ii) If there is a sequence $\{X_n\} \subset \mathfrak{A}$ such that $\mu(X_n) < +\infty$, $n = 1, 2, \ldots$, and $X = \bigcup_{n=1}^{\infty} X_n$, then $\mu(A) = \nu(A)$ for each $A \in \mathfrak{M}$.

Hint: Modify the sequence $\{X_n\}$ so that the sets $X_1, X_2, \ldots$ are disjoint. Observe that for $n = 1, 2, \ldots$ the algebras $\mathfrak{A}_n = \{A \in \mathfrak{A} : A \subset X_n$ in X_n generate the σ-algebras $\mathfrak{M}_n = \{A \in \mathfrak{M} : A \subset X_n\}$. Then use (i).

(16-2) Let $X = \mathbf{R}$, and let $\mathfrak{A}$ be the algebra in X generated by intervals $[a, +\infty)$, $a \in \mathbf{R}$ [see (7-4)(iii)]. Furthermore, let P and Q be disjoint countable dense subsets of $\mathbf{R}$, and let μ_1 and ν_1 be the counting measures on exp P and exp Q, respectively. Show that

(i) The σ-algebra generated by $\mathfrak{A}$ is the Borel σ-algebra $\mathfrak{B}$ in $\mathbf{R}$.

(ii) If $\mu(A) = \mu_1(A \cap P)$ and $\nu(A) = \nu_1(A \cap Q)$ for each $A \in \mathfrak{B}$, then μ and ν are σ-finite measures on $\mathfrak{B}$.

(iii) $\mu(A) = \nu(A)$ for each $A \in \mathfrak{A}$ and yet $\mu \neq \nu$.

In Exercises (16-3) through (16-10) we shall assume that $(X_i, \mathfrak{M}_i, \mu_i)$, $i = 1, 2$, are measure spaces and that $(X, \mathfrak{J}, \iota)$ is their product. We shall use freely the notation established in this chapter.

Let $\mathfrak{M}$ be the σ-algebra in X generated by $\mathfrak{R}$ [see (7.3)]. It follows from (16.4) that $\mathfrak{M} \subset \mathfrak{I}$. We shall denote by μ the restriction of ι to $\mathfrak{M}$. We note that in some literature (e.g., in Refs. 10 and 13) the symbols $\mathfrak{M}_1 \times \mathfrak{M}_2$ and $\mu_1 \times \mu_2$ are used to denote $\mathfrak{M}$ and μ, respectively.

(16-3)$^+$ Let $\mathfrak{F}_\mu$ and I_μ denote, respectively, the fundamental system and the integral induced by the measure μ, and similarly, let $\mathfrak{F}_\iota$ and I_ι be induced by the measure ι [see (12.10)]. Show that

(i) $\mathfrak{F} \subset \mathfrak{F}_\mu \subset \mathfrak{F}_\iota$, the integrals I and I_μ coincide on $\mathfrak{F}$, and the integrals I_μ and I_ι coincide on $\mathfrak{F}_\mu$.

(ii) The measure space $(X, \mathfrak{I}, \iota)$ is induced by the measure μ. Hint: By (12.18) the integrals I and I_ι coincide on the family of integrable functions. Use this, (i), and (6.22).

(iii) The measure space $(X, \mathfrak{I}, \iota)$ is obtained from the measure space $(X, \mathfrak{M}, \mu)$ by the completion and saturation (in this order) of the measure μ [see (8-9) and (8-10)]. Hint: Use (ii) and (12-12).

(16-4)$^{*+}$ Show that

(i) ι is σ-finite iff μ is. Hint: Use (16-3)(iii).

(ii) If μ_1 and μ_2 are σ-finite, then so is μ.

(iii) If $A \in \mathfrak{I}$ and $\iota(A) < +\infty$, then there are μ_i-σ-finite sets $A_i \in \mathfrak{M}_i$, $i = 1, 2$, such that $A \subset A_1 \times A_2$. Hint: If $f \in \mathfrak{F}$, then there is a $B_1 \times B_2 \in \mathfrak{R}_o$ for which $\{x \in X : f(x) \neq 0\} \subset B_1 \times B_2$. It follows that if $f \in \mathfrak{F}_+^*$, then there are μ_i-σ-finite sets $A_i \in \mathfrak{M}_i$, $i = 1, 2$, such that $\{x \in X : f(x) \neq 0\} \subset A_1 \times A_2$.

(iv) Let $A_1 \times A_2 \in \mathfrak{R}$. If either A_1 is not μ_1-σ-finite or A_2 is not μ_2-σ-finite, then $\mu(A_1 \times A_2) = +\infty$.

(v) If $A_i \in \mathfrak{M}_i$, $i = 1, 2$, are μ_i-σ-finite, then

$$\mu(A_1 \times A_2) = \mu_1(A_1)\mu_2(A_2)$$

Hint: Use (16.8).

(vi) If $X_i \neq \emptyset$, $i = 1, 2$, then μ is σ-finite iff μ_1 and μ_2 are. Hint: Use (ii) and (iii).

(16-5)[+] Let μ_i, i = 1, 2, be σ-finite. Show that

(i) If $A \times B \in \mathfrak{R}$, then $\mu(A \times B) = \mu_1(A)\mu_2(B)$. Hint: Use (16-4)(v).

(ii) If ν is a measure on $\mathfrak{M}$ such that $\nu(A \times B) = \mu_1(A)\mu_2(B)$ for each $A \times B \in \mathfrak{R}$, then $\nu = \mu$. Hint: Observe that the disjoint unions of sets from $\mathfrak{R}$ form an algebra in X, and apply (16-1)(ii).

(16-6)[+] Let $A \in \mathfrak{M}$, and let f be an $\mathfrak{M}$-measurable function on X. Show that

(i) $A_x \in \mathfrak{M}_2$ for each $x \in X_1$ and $A^y \in \mathfrak{M}_1$ for each $y \in X_2$. Hint: Show that the family of all sets $A \subset X$ with this property is a σ-algebra in X containing $\mathfrak{R}$. Then apply (7.3).

(ii) f_x is $\mathfrak{M}_2$-measurable for each $x \in X_1$, and f^y is $\mathfrak{M}_1$-measurable for each $y \in X_2$. Hint: Use (i) to show that this is true for a simple function f and apply (7.29).

In the following exercises we shall use the n-dimensional Lebesgue measure λ_n, $n \geq 1$ an integer, which was defined in (10-7).

(16-7)[+] Let $(X_1, \mathfrak{M}_1, \mu_1) = (\mathbf{R}^r, \Lambda_r, \lambda_r)$ and $(X_2, \mathfrak{M}_2, \mu_2) = (\mathbf{R}^s, \Lambda_s, \lambda_s)$, where r and s are positive integers. Prove

(i) $(X, \mathfrak{I}, \iota) = (\mathbf{R}^{r+s}, \Lambda_{r+s}, \lambda_{r+s})$.

(ii) If $M \subset \mathbf{R}$ is a Lebesgue nonmeasurable set [see (10-8)] and M^r is the r-fold cartesian product of M, then $M^r \notin \Lambda_r$. Hint: Using (16.8) proceed by induction.

(iii) $\mathfrak{M} \subsetneqq \mathfrak{I}$ and the measure μ is not complete. Hint: Choose a set $A \subset \mathbf{R}^r$ so that $A \notin \Lambda_r$ and a set $B \in \Lambda_s$ so that $B \neq \emptyset$ and $\lambda_s(B) = 0$. Using (16-6)(i), observe that $A \times B \notin \mathfrak{M}$. Since $A \times B \subset \mathbf{R}^r \times B$ and $\lambda_{r+s}(\mathbf{R}^r \times B) = 0$ [see (16-5)(i)], we have $A \times B \in \mathfrak{I}$.

(16-8) Let $X_1 = X_2 = \mathbf{R}$, $\mathfrak{M}_1 = \mathfrak{M}_2 = \mathfrak{B}$, where $\mathfrak{B}$ is the Borel σ-algebra in $\mathbf{R}$, and let $\mu_1 = \mu_2$ be the restriction of the Lebesgue measure λ to $\mathfrak{B}$. Furthermore, let A be a subset of the Cantor discontinuum which is not a Borel set [see (11-1)(ii)], and let $C = A \times \mathbf{R}$. Show that $C \in \mathfrak{I}$, and for no $y \in \mathbf{R}$, $C^y \in \mathfrak{B}$. Conclude that the completeness of the measures μ_1 and μ_2 is essential for the validity of (16.6) through (16.8).

However, the completeness of μ_1 and μ_2 is not necessary if we replace $\mathfrak{I}$ by $\mathfrak{M}$ in (16.6) through (16.8). Use (16-6) to prove it.

(16-9) Let $(X_1, \mathfrak{M}_1, \mu_1)$ be an arbitrary measure space with a σ-finite measure μ_1, and let $(X_2, \mathfrak{M}_2, \mu_2) = (\mathbf{R}, \Lambda, \lambda)$. If f is an $\mathfrak{M}_1$-measurable function on a set $A \in \mathfrak{M}_1$, show that

(i) If $M_f = \{(x,y) \in A \times \mathbf{R} : 0 \leq y < f(x)\}$, then $M_f \in \mathfrak{M}$ and $\mu(M_f) = \int_A f^+ \, d\mu_1$. Hint: Show that $M_f \in \mathfrak{M}$ whenever f is a simple function [use (16-5)], and then apply (7.29). By (16.5) and (16.8), $\mu(M_f) = \int_A f^+ \, d\iota_1$. Since f is $\mathfrak{M}_1$-measurable, $\int_A f^+ \, d\iota_1 = \int_A f^+ \, d\mu_1$.

(ii) If f is finite and $N_f = \{(x,y) \in A \times \mathbf{R} : 0 \leq y \leq f(x)\}$, then $N_f \in \mathfrak{M}$ and $\mu(N_f) = \mu(M_f)$.

(iii) If f is finite and $\Gamma_f = \{(x,y) \in \mathbf{R} : f(x) = y\}$ is the graph of f, then $\Gamma_f \in \mathfrak{M}$ and $\mu(\Gamma_f) = 0$.

(iv) For an integer $n \geq 0$, let S^n be the unit n-sphere and B^n be the unit n-ball, i.e.,

$$S^n = \{(x_0, \ldots, x_n) \in \mathbf{R}^{n+1} : \sum_{i=0}^{n} x_i^2 = 1\}$$

$$B^n = \{(x_1, \ldots, x_n) \in \mathbf{R}^n : \sum_{i=1}^{n} x_i^2 \leq 1\}$$

Then $S^n \in \Lambda_{n+1}$, $B^n \in \Lambda_n$, and $\lambda_{n+1}(S^n) = 0$. Hint: Using (ii) and (iii), proceed by induction.

(16-10)* We shall consider the J. von Neumann extension λ'_n of λ_n defined in (10-14). Let $(X_1, \mathfrak{M}_1, \mu_1) = (\mathbf{R}^r, \Lambda'_r, \lambda'_r)$ and $(X_2, \mathfrak{M}_2, \mu_2) = (\mathbf{R}^s, \Lambda'_s, \lambda'_s)$, where r and s are positive integers. Show that

(i) $\Lambda_{r+s} \subset \mathfrak{I}$ and $\iota(A) = \lambda_{r+s}(A)$ for each $A \in \Lambda_{r+s}$. Hint: Use (16-7) and (6-6)(ii).

(ii) If $C \subset \mathbf{R}^{r+s}$ and $\|C\| < c$, then $C \in \mathfrak{I}$ and $\iota(C) = 0$. Hint: Observe that there are $A \subset \mathbf{R}^r$ and $B \subset \mathbf{R}^s$ such that $\|A\| < c$, $\|B\| < c$, and $C \subset A \times B$.

(iii) $\Lambda'_{r+s} \subset \mathfrak{I}$ and $\iota(A) = \lambda'_{r+s}(A)$ for each $A \in \Lambda'_{r+s}$. Hint: If $A \in \Lambda'_{r+s}$, then there is a $B \in \Lambda_{r+s}$ such that $\|(A - B) \cup (B - A)\| < c$. Observe that $A = (A - B) \cup [B - (B - A)]$, and apply (i) and (ii).

(iv) If $\Lambda_r \subsetneqq \Lambda'_r$, then $\Lambda'_{r+s} \subsetneqq \mathfrak{I}$. Hint: Choose $A' \in \Lambda'_r - \Lambda_r$, and let $C' = A' \times \mathbf{R}^s$. Observe that if $C \subset \mathbf{R}^{r+s}$ is such that $\|(C - C') \cup (C' - C)\| < c$, then $\|\{y \in \mathbf{R}^s : C^y \in \Lambda_r\}\| < c$. Using (16.8) conclude that $C' \notin \Lambda'_{r+s}$.

(16-11) Let f and g be λ-summable functions on $\mathbf{R}$, and let $h(x,y) = f(x)g(y - x)$ for each $(x,y) \in \mathbf{R}^2$. Show that h is λ_2-integrable on $\mathbf{R}^2$ and that

$$\int_{\mathbf{R}^2} h \, d\lambda_2 = \left(\int_{\mathbf{R}} f \, d\lambda\right)\left(\int_{\mathbf{R}} g \, d\lambda\right)$$

(16-12) Let $A = [-1,1] \times [-1,1]$, and for $(x,y) \in A$, let $f(x,y) = xy(x^2 + y^2)^{-2}$ if $(x,y) \neq (0,0)$, and $f(0,0) = 0$. Show that

$$\int_{-1}^{1}\left(\int_{-1}^{1} f(x,y)\, dx\right) dy = \int_{-1}^{1}\left(\int_{-1}^{1} f(x,y)\, dy\right) dx = 0$$

and that the integral $\int_A f \, d\lambda_2$ does not exist. Hint: Show that

$$\int_A f^+ \, d\lambda_2 = \int_A f^- \, d\lambda_2 = +\infty$$

(16-13) Let $X = [0,1] \times [0,1]$. If $\mathfrak{C}$ is the family of all compact sets $F \subset X$ for which $\lambda_2(F) > 0$, it follows from (7-11)(ii) that $\|\mathfrak{C}\| = c$. Thus there is a one-to-one map $\alpha \mapsto F_\alpha$ from the set $\{\alpha : \alpha < \zeta\}$ onto $\mathfrak{C}$ [see Chapter 1, Section A]. Construct a set $A \subset X$ by transfinite induction as follows: Choose $(x_0, y_0) \in F_0$ and suppose that $(x_\alpha, y_\alpha) \in F_\alpha$ has been already chosen for all $\alpha < \beta < \zeta$ so that if $S = \{(x_\alpha, y_\alpha) : \alpha < \beta\}$, then S_x and S^x contain at most one element for each $x \in [0,1]$. Let $B = \{x \in [0,1] : \lambda_1[(F_\beta)_x] > 0\}$. Because $\lambda_2(F_\beta) > 0$, by (16.8) we have $\lambda_1(B) > 0$. Since $\|S\| < c$, it follows from (10-13)(ii) that there is an $x_\beta \in B$ such that $x_\beta \neq x_\alpha$ for each $\alpha < \beta$. By the definition of B,

$\lambda_1[(F_\beta)_{x_\beta}] > 0$, and so again there is a $y_\beta \in (F_\beta)_{x_\beta}$ such that $y_\beta \neq y_\alpha$ for each $\alpha < \beta$. We set $A = \{(x_\alpha, y_\alpha) : \alpha < \zeta\}$.

Show that

(i) For each $x \in [0,1]$, A_x and A^x contain at most one element.

(ii) $A \cap F \neq \emptyset$ for each $F \in \mathfrak{C}$.

(iii) If $A \in \Lambda_2$, then $\lambda_2(A) = 0$. Hint: Use (16.8).

(iv) If $A \in \Lambda_2$, then $\lambda_2(X - A) = 0$. Hint: Observe that λ_2 is a regular measure and use (ii) and (9.2).

Conclude from (iii) and (iv) that $A \notin \Lambda_2$.

(16-14) Let $\alpha \mapsto x_\alpha$ be a one-to-one map from the set $\{\alpha : \alpha < \zeta\}$ onto $[0,1]$ (see Chapter 1, Section A). Let $X = [0,1] \times [0,1]$, and let

$$A = \{(x_\alpha, x_\beta) \in X : \alpha < \beta < \zeta\}$$

Show that $\|A^x\| < c$ and $\|(X - A)_x\| < c$ for each $x \in [0,1]$. Use (16.8) and (16-10)(iii) to conclude that $A \notin \Lambda_2'$ [see (10-14)].

17. PRODUCT MEASURES IN A TOPOLOGICAL SPACE

The product of measures defined in Chapter 16 has a serious shortcoming when applied to measures in topological spaces. Namely, the product of two regular measures may not be a regular measure [see (17-10)]. For this reason, in the topological situation we shall combine the results of Chapters 13, 15, and 16, and we shall define a new product measure which is regular whenever the factor measures are.

Throughout this chapter we shall keep the assumptions, notation, and terminology of Chapter 16. In addition, we shall assume the following:

(i) X_1 and X_2 are locally compact Hausdorff spaces, and the space $X = X_1 \times X_2$ has the product topology

(ii) If F is a compact G_δ subset of X_i, then $F \in \mathfrak{M}_i$ and $\mu_i(F) < +\infty$, $i = 1, 2$.

Note that according to assumption (i), the space X is a locally compact Hausdorff space. Also note that assumption (ii) is satisfied whenever the measures μ_1 and μ_2 are regular.

By $\mathfrak{F}$ and $\mathfrak{G}$ we shall denote the families of all compact and all open subsets of X, respectively. The family of all σ-compact subsets of X is denoted by $\mathfrak{F}_\sigma$, and the family of all G_δ subsets of X is denoted by $\mathfrak{G}_\delta$. For $i = 1, 2$, we shall denote by $\mathfrak{F}_i$, $\mathfrak{G}_i$, $\mathfrak{F}_{i,\sigma}$, and $\mathfrak{G}_{i,\delta}$ the corresponding families of subsets of X_i.

(17.1) Lemma. The family $\mathfrak{F} \cap \mathfrak{G}_\delta$ is contained in the σ-algebra in X generated by the collection

$$\{F_1 \times F_2 : F_i \in \mathfrak{F}_i \cap \mathfrak{G}_{i,\delta}, i = 1, 2\}$$

Proof. Let $F \in \mathfrak{J} \cap \mathfrak{G}_\delta$, and let $G \in \mathfrak{G}$ contain F. By (13.3), each $x \in F$ has a neighborhood $W = V_1 \times V_2$ such that $W \subset G$ and $V_i \in \mathfrak{J}_i \cap \mathfrak{G}_{i,\delta}$, i = 1, 2. Because F is compact, it is covered by finitely many of these neighborhoods, say $W_1, \ldots, W_n$. Clearly, $F \subset \bigcup_{j=1}^n W_j \subset G$, and since F is G_δ, the lemma follows. ///

(17.2) Corollary. $\mathfrak{J} \cap \mathfrak{G}_\delta \subset \mathfrak{J}$.

This corollary follows from assumption (ii), (17.1), and (16.4).

(17.3) Corollary. $C_o(X_i) \subset \mathfrak{L}(X_i)$, i = 1, 2, and $C_o(X) \subset \mathfrak{L}(X)$.

Using assumption (ii) and (17.2), the proof of Corollary (17.3) is the same as the first part of the proof of Proposition (13.6).

For i = 1, 2, denote by J_i and $(X_i, \mathfrak{N}_i, \nu_i)$, respectively, the integral and the measure space topologically induced by the measure μ_i [see (13.11)]. Similarly, let J and $(X, \mathfrak{N}, \nu)$ be, respectively, the integral and the measure space topologically induced by the product measure ι.

Recall that J_i is a Bourbaki integral on $C_o(X_i)$ which is the restriction to $C_o(X_i)$ of the integral I_i induced by the measure μ_i, i = 1, 2. Similarly, J is a Bourbaki integral on $C_o(X)$ which is the restriction to $C_o(X)$ of the product integral I [which induces the product measure ι and which by (12.18) is also induced by ι]. The measure spaces $(X_1, \mathfrak{N}_1, \nu_1)$, $(X_2, \mathfrak{N}_2, \nu_2)$, and $(X, \mathfrak{N}, \nu)$ are induced by the integrals J_1, J_2, and J, respectively. Consequently, the measures ν_1, ν_2, and ν are regular complete and saturated [see (10.18) and (10.19)].

(17.4) Definition. The measure space $(X, \mathfrak{N}, \nu)$ topologically induced by the product measure ι is called the topological product of the measure spaces $(X_1, \mathfrak{M}_1, \mu_1)$ and $(X_2, \mathfrak{M}_2, \mu_2)$.

Notice that the measure ν is always regular complete and saturated, regardless whether the measures μ_1 and μ_2 are. Let $(X_i, \mathfrak{I}_i, \iota_i)$ be the measure space induced by the measure μ_i, i = 1, 2 [see (12.10)]. Then it follows from (16.5) that $(X, \mathfrak{N}, \nu)$ is also the topological product of $(X_1, \mathfrak{I}_1, \iota_1)$ and $(X_2, \mathfrak{I}_2, \iota_2)$.

The use of the term "product" is justified by the following proposition.

(17.5) Proposition. Let $g \in C_o(X_1)$ and $h \in C_o(X_2)$. If $f = g \otimes h$, then

$$\int_X f\, d\nu = \left(\int_{X_1} g\, d\mu_1\right)\left(\int_{X_2} h\, d\mu_2\right)$$

Proof. From (17.3), (16.5), and (16.6) we obtain

$$\int_X f\, d\nu = \int_X f\, d\iota = \int_{X_1}\left(\int_{X_2} f_x\, d\iota_2\right) d\iota_1$$

$$= \left(\int_{X_1} g\, d\iota_1\right)\left(\int_{X_2} h\, d\iota_2\right) = \left(\int_{X_1} g\, d\mu_1\right)\left(\int_{X_2} h\, d\mu_2\right) \quad ///$$

In the previous proof we could not apply Theorme (16.6) directly to the measures μ_1 and μ_2 , since they may be incomplete.

We shall show that the measure ν is the only regular complete and saturated measure in X for which Proposition (17.5) holds. To do this we shall need a certain version of the Stone-Weierstrass theorem, which we shall present without proof. For a proof we refer the reader to Ref. 17, Section 4. The theorem below is obtained by applying Theorem 4E of Ref. 17, Section 4, p. 9, to the one-point compactification of the space Z.

(17.6) Theorem (Stone-Weierstrass). Let Z be a locally compact Hausdorff space, and let $\mathfrak{A} \subset C_o(Z)$ satisfy the following conditions:

(i) If f and g belong to $\mathfrak{A}$, then so does $f + g$ and fg

(ii) If $f \in \mathfrak{A}$ and $a \in \mathbf{R}$, then $af \in \mathfrak{A}$

(iii) For each $z \in Z$, there is an $f \in \mathfrak{A}$ such that $f(z) \neq 0$

(iv) For each pair of distinct points $z_i \in Z$, $i = 1, 2$, there is an $f \in \mathfrak{A}$ such that $f(z_1) \neq f(z_2)$.

Then for every $f \in C_o(Z)$ and every $\epsilon > 0$ there is a $g \in \mathfrak{A}$ such that

$$|f(z) - g(z)| < \epsilon$$

for each $z \in Z$.

(17.7) Lemma. Let $f \in C_o(X)$. We can find an $A \in \mathfrak{J}$ such that for every $\epsilon > 0$, there is a $g \in C_o(X_1) \otimes C_o(X_2)$ for which

$$|f - g| \leq \epsilon\chi_A$$

Proof. Clearly, there are open sets $U \subset X_1$ and $V \subset X_2$ such that U^- and V^- are compact and supp $f \subset U \times V$. Set $A = U^- \times V^-$, and choose $\epsilon > 0$. It is easy to see that if we let $Z = U \times V$, then the family $\mathfrak{A} = \mathfrak{C}_o(U) \otimes \mathfrak{C}_o(V)$ satisfies the conditions (i) through (iv) of Theorem (17.6). Thus there are functions $h_j \in \mathfrak{C}_o(U)$ and $k_j \in \mathfrak{C}_o(V)$, $j = 1, \ldots, n$, such that

$$|f(x,y) - \sum_{j=1}^{n} h_j(x)g_j(y)| < \epsilon$$

for each $(x,y) \in U \times V$. Let $h_j^\wedge$ and $k_j^\wedge$, $j = 1, \ldots, n$, be the normal extensions [see the paragraph preceding (14.1)] of h_j and k_j, respectively, and let $g = \Sigma_{j=1}^{n} h_j^\wedge \otimes k_j^\wedge$. Then $g \in \mathfrak{C}_o(X_1) \otimes \mathfrak{C}_o(X_2)$ and

$$|f - g| \leq \epsilon\chi_A \quad ///$$

(17.8) Theorem. Let $(X,\mathfrak{M},\mu)$ be a measure space with a regular complete and saturated measure μ, and let

$$\int_X (g \otimes h)\, d\mu = \left(\int_{X_1} g\, d\mu_1\right)\left(\int_{X_2} h\, d\mu_2\right)$$

for each $g \in \mathfrak{C}_o(X_1)$ and each $h \in \mathfrak{C}_o(X_2)$. Then $(X,\mathfrak{M},\mu) = (X,\mathfrak{A},\nu)$.

Proof. Let $f \in \mathfrak{C}_o(X)$, and let $A \in \mathfrak{J}$ be the set from Lemma (17.7). Choose $\epsilon > 0$. By (17.7), there is a $g \in \mathfrak{C}_o(X_1) \otimes \mathfrak{C}_o(X_2)$ such that

$$|f - g| \leq \epsilon\chi_A$$

It follows from (17.5) that

$$\int_X g\, d\mu = \int_X g\, d\nu$$

and so

$$\left|\int_X f\, d\mu - \int_X f\, d\nu\right| \leq \left|\int_X f\, d\mu - \int_X g\, d\mu\right| + \left|\int_X g\, d\nu - \int_X f\, d\nu\right|$$

$$\leq \int_X |f - g|\, d\mu + \int_X |g - f|\, d\nu \leq \epsilon[\mu(A) + \nu(A)]$$

Since A is compact, $\mu(A)$ and $\nu(A)$ are real numbers. Thus by the arbitrariness of ϵ,

$$\int_X f\,d\mu = \int_X f\,d\nu$$

The theorem follows from (13.9). ///

(17.9) Corollary. The topological product of the measure spaces $(X_1, \mathfrak{N}_1, \nu_1)$ and $(X_2, \mathfrak{N}_2, \nu_2)$ is the measure space $(X, \mathfrak{N}, \nu)$.

Proof. By (17.5) we have

$$\int_X (g \otimes h)\,d\nu = \left(\int_{X_1} g\,d\mu_1\right)\left(\int_{X_2} h\,d\mu_2\right) = \left(\int_{X_1} g\,d\nu_1\right)\left(\int_{X_2} h\,d\nu_2\right)$$

for each $g \in \mathfrak{C}_o(X_1)$ and each $h \in \mathfrak{C}_o(X_2)$. Thus the corollary follows from Theorem (17.8) applied to the measures ν_1 and ν_2. ///

We shall prove that the integral J is the product of the integrals J_1 and J_2.

(17.10) Lemma. Let $f \in \mathfrak{C}_o(X)$. Then we can find a set $A \in \mathfrak{J}_1 \cap \mathfrak{G}_{1,\delta}$ with the following property: To every $\epsilon > 0$ and every $y \in X_2$ there is a neighborhood V of y such that

$$|f^y - f^\eta| \le \epsilon \chi_A$$

for each $\eta \in V$.

Proof. Since the set

$$B = \{x \in X_1 : (\text{supp } f)_x \ne \emptyset\}$$

is the projection of supp f to X_1, it is compact. By (13.3) we can find an $A \in \mathfrak{J}_1 \cap \mathfrak{G}_{1,\delta}$ which contains B. Choose $\epsilon > 0$ and $y \in X_2$. To every $x \in A$ there are neighborhoods U_x of x and V_x of y such that

$$|f(x,y) - f(\xi,\eta)| < \frac{\epsilon}{2}$$

for all $(\xi,\eta) \in U_x \times V_x$. Because A is compact, we can find $x_1, \ldots, x_n$ in A so that $A \subset \bigcup_{i=1}^n U_{x_i}$. Clearly, $V = \bigcap_{i=1}^n V_{x_i}$ is a neighborhood of y. Choose $\eta \in V$. If $x \in A$, then $x \in U_{x_i}$ for some integer i, $1 \le i \le n$, and we have

$$|f(x,y) - f(x,\eta)| \le |f(x,y) - f(x_i,y)| + |f(x_i,y) - f(x,\eta)| < \epsilon = \epsilon \chi_A(x)$$

If $x \in X_1 - A$, then

$$|f(x,y) - f(x,\eta)| = 0 = \epsilon \chi_A(x)$$

The lemma follows. ///

(17.11) Proposition. The system $\mathfrak{C}_o(X)$ is an ample product system with respect to $(\mathfrak{C}_o(X_1), J_1)$ and $(\mathfrak{C}_o(X_2), J_2)$. Moreover the integral J on $\mathfrak{C}_o(X)$ is the product of the integrals J_1 on $\mathfrak{C}_o(X_1)$ and J_2 on $\mathfrak{C}_o(X_2)$.

Proof. Clearly $\mathfrak{C}_o(X_1) \otimes \mathfrak{C}_o(X_2) \subset \mathfrak{C}_o(X)$. Let $f \in \mathfrak{C}_o(X)$.

(a) It is obvious that $f_x \in \mathfrak{C}_o(X_2)$ for each $x \in X_1$ and $f^y \in \mathfrak{C}_o(X_1)$ for each $y \in X_2$.

(b) Let $A \in \mathfrak{F}_1 \cap \mathfrak{G}_{1,\delta}$ be the set from Lemma (17.10). By assumption (ii) from the beginning of this chapter, $A \in \mathfrak{M}_1$ and $\mu_1(A) < +\infty$. Choose $\epsilon > 0$ and $y \in X_2$. According to (7.10), there is a neighborhood V of y such that

$$|f^y - f^\eta| \le \frac{\epsilon}{1 + \mu_1(A)} \chi_A$$

for each $\eta \in V$. Using (12.19) we obtain

$$|J_1(f^y) - J_1(f^\eta)| = \left|\int_{X_1} f^y \, d\mu_1 - \int_{X_1} f^\eta \, d\mu_1\right| \le \int_{X_1} |f^y - f^\eta| \, d\mu_1 \le \frac{\epsilon}{1 + \mu_1(A)} \mu_1(A) < \epsilon$$

for each $\eta \in V$. It follows that $J_1 f \in \mathfrak{C}_o(X_2)$ and by symmetry $J_2 f \in \mathfrak{C}_o(X_1)$.

(c) From (a), (b), (17.3), (16.5), and (16.6) we obtain

$$Jf = \int_X f \, d\iota = \int_{X_1} \left(\int_{X_2} f_x \, d\iota_2 \right) d\iota_1 = J_1(J_2 f)$$

and similarly, $Jf = J_2(J_1 f)$. ///

Since, in general, there is little connection between the measures ν and ι, there is even less connection between the measure ν and the measures μ_1 and μ_2. In order to make some use of Proposition (17.11) we must therefore make an additional assumption about the measures μ_1 and μ_2.

Throughout the remainder of this chapter we shall assume that μ_1 and μ_2 are <u>regular</u> measures.

<u>(17.12)</u> <u>Theorem</u>. Let μ_1 and μ_2 be complete measures, and let f be a ν-summable function on X. Then

$$\int_X f\,d\nu = \int_{X_1}\left(\int_{X_2} f_x\,d\mu_2\right)d\mu_1 = \int_{X_2}\left(\int_{X_1} f^y\,d\mu_1\right)d\mu_2$$

<u>Proof</u>. The proof of this theorem is quite similar to that of Theorem (16.6). By (12.22) the function f is J-summable and $\int_X f\,d\nu = Jf$. From (13.1) and (13.13) we obtain that $(X_i, \mathfrak{N}_i, \nu_i) = (X_i, \mathfrak{I}_i, \iota_i)$, i = 1, 2. Thus according to (12.18), for i = 1, 2, the integral J_i and the measure space $(X_i, \mathfrak{I}_i, \iota_i)$ are mutually inducing each other. Since the measures μ_1 and μ_2 are complete, it follows from (12.19) and (12.20) that $\mathfrak{L}(X_i)$ consists of all J_i-summable functions and that $J_i h = \int_{X_i} h\,d\mu_i$ for each $h \in \mathfrak{L}(X_i)$. Moreover, by (12.14) and (12.16) the null ideals $\mathfrak{M}_{i,o}$ and $\mathfrak{I}_{i,o}$ coincide [see (8.7)]. Hence "μ_i-almost all" means the same as "ι_i-almost all," i = 1, 2. The theorem follows from (17.11) and (15.7). ///

The next theorem and its corollary follow from Theorem (17.12) in the exactly same manner in which Theorem (16.7) and Corollary (16.8) followed from Theorem (16.6).

<u>(17.13)</u> <u>Theorem</u>. Let μ_1 and μ_2 be complete measures. If $A \in \mathfrak{N}$ is a ν-σ-finite set and if f is a ν-integrable function on A, then

$$\int_A f\,d\nu = \int_{X_1}\left(\int_{A_x} f_x\,d\mu_2\right)d\mu_1 = \int_{X_2}\left(\int_{A^y} f^y\,d\mu_1\right)d\mu_2$$

<u>(17.14)</u> <u>Corollary</u>. Let μ_1 and μ_2 be complete measures. If $A \in \mathfrak{N}$ is a ν-σ-finite set, then

$$\nu(A) = \int_{X_1} \mu_2(A_x)\, d\mu_1 = \int_{X_2} \mu_1(A^y)\, d\mu_2$$

Even if the measures μ_1 and μ_2 are regular, the relationship between their product ι and their topological product ν is not simple.

(17.15) Theorem. $\mathfrak{I} \subset \mathfrak{N}$ and $\nu(A) \leq \iota(A)$ for each $A \in \mathfrak{I}$. Moreover, if $A \in \mathfrak{I}$ is ι-σ-finite, then $\nu(A) = \iota(A)$.

Proof. For the purpose of this proof only we shall denote by $\mathfrak{L}_I$, $\bar{\mathfrak{L}}_I$, $\mathfrak{L}_J$, and $\bar{\mathfrak{L}}_J$ the families of functions on X which are, respectively, I-summable, I-integrable, J-summable, and J-integrable.

(a) If $A \in \mathfrak{K}_0$, then by the regularity of μ_1 and μ_2, (17.11), (15.9), and (16.4), we have $A \in \mathfrak{N}$ and $\nu(A) = \iota(A)$. It follows that $\mathfrak{F} \subset \mathfrak{L}_J$ [see (16.2)] and $Jf = If$ for each $f \in \mathfrak{F}$. Because I is a Daniell integral on $\mathfrak{F}$, also $\bar{\mathfrak{F}} \subset \bar{\mathfrak{L}}_J$ and $Jf = If$ for each $f \in \bar{\mathfrak{F}}$. Therefore,

$$\underline{I}f \leq \underline{J}f \leq \bar{J}f \leq \bar{I}f$$

for every function f on X.

(b) By (a), $\mathfrak{L}_I \subset \mathfrak{L}_J$ and $Jf = If$ for each $f \in \mathfrak{L}_I$. Hence, if $A \in \mathfrak{I}$ and $\iota(A) < +\infty$, then $A \in \mathfrak{N}$ and $\nu(A) = \iota(A)$. Moreover, (8.5) implies that this remains correct if $A \in \mathfrak{I}$ is ι-σ-finite.

(c) Since $C_0(X) \subset \mathfrak{L}_I$ [see (17.3)], it follows from (b) and (6.9) that $\bar{\mathfrak{L}}_I \subset \bar{\mathfrak{L}}_J$. Therefore $\mathfrak{I} \subset \mathfrak{N}$. By (a), $\nu(A) \leq \iota(A)$ for each $A \in \mathfrak{I}$. ///

(17.16) Corollary. Let the measures μ_1 and μ_2 be σ-finite. Then $(X, \mathfrak{N}, \nu) = (X, \mathfrak{I}, \iota)$ iff $\mathfrak{G} \subset \mathfrak{I}$.

Proof. Let $\mathfrak{G} \subset \mathfrak{I}$. Since μ_1 and μ_2 are σ-finite, we obtain from (16.4) that ι is also σ-finite. Thus by (17.15), ι is a regular measure, and it follows from (10.18) and (9.10) that $(X, \mathfrak{N}, \nu) = (X, \mathfrak{I}, \iota)$.

Since ν is a regular measure, the converse is trivial. ///

We note that if the σ-finiteness of μ_1 and μ_2 is not assumed, the previous corollary is generally false [see (17.10)]. The following question is an unsolved problem.

Question: Is the product (nontopological) of two regular σ-finite measures also regular?

We shall give the affirmative answer in an important special case.

(17.17) Proposition. Let the spaces X_1 and X_2 be metrizable and σ-compact. Then $(X, \mathfrak{N}, \nu) = (X, \mathfrak{I}, \iota)$.

Proof. Clearly, the space X is metrizable and σ-compact. From this we easily obtain that $\mathfrak{F} \subset \mathfrak{G}_\delta$ and $\mathfrak{G} \subset \mathfrak{F}_\sigma$. Thus by (17.2), $\mathfrak{G} \subset \mathfrak{I}$ and the proposition follows from (17.16). ///

Exercises

Throughout these exercises we shall use the notation established in Chapters 16 and 17. We shall assume that for i = 1, 2, $(X_i, \mathfrak{M}_i, \mu_i)$ is a measure space such that X_i is a locally compact Hausdorff space. $\mathfrak{F}_i \cap \mathfrak{G}_{i,\delta} \subset \mathfrak{M}_i$, and $\mu_i(F) < +\infty$ for each $F \in \mathfrak{F}_i \cap \mathfrak{G}_{i,\delta}$.

In exercises (17-1) through (17-5) we shall show how the measure space $(X, \mathfrak{N}, \nu)$ can be induced by a suitably defined volume. For the terminology concerning the volume spaces we refer to the exercises (13-7) through (13-12).

(17-1) Let $A \times B = \bigcup_{i=1}^n A_i \times B_i$, where $A_i \times B_i$, $i = 1, \ldots, n$, are disjoint sets from $\mathfrak{R}_0$ (recall from Chapter 16 that $\mathfrak{R}_0$ is the family of all measurable rectangles with finite measure). Without using (16.4), show that

$$\mu_1(A)\mu_2(B) = \sum_{i=1}^n \mu_1(A_i)\mu_2(B_i)$$

Hint: Let $\{A'_j\} \subset \mathfrak{M}_1$ and $\{B'_k\} \subset \mathfrak{M}_2$ be finite disjoint strong refinements of $\{A_i\}_{i=1}^n$ and $\{B_i\}_{i=1}^n$, respectively [see (12.6)]. Observe that if $C \times D$ is a union of some $A'_j \times B'_k$, then

$$\mu_1(C)\mu_2(D) = \Sigma\{\mu_1(A'_j)\mu_2(B'_k) : A'_j \times B'_k \subset C \times D\}$$

(17-2) Let $\{A_j \times B_j\}_{j=1}^n \subset \mathfrak{R}_0$ and $\{A'_k \times B'_k\}_{k=1}^m \subset \mathfrak{R}_0$ be two disjoint families such that

$$\bigcup_{j=1}^{n} (A_j \times B_j) = \bigcup_{k=1}^{m} (A_k' \times B_k')$$

Without using (16.4), show that

$$\sum_{j=1}^{n} \mu_1(A_j)\mu_2(B_j) = \sum_{k=1}^{n} \mu_1(A_k')\mu_2(B_k')$$

Hint: Use (16.1) to find a strong disjoint refinement of

$$\{A_j \times B_j \,,\, A_k' \times B_k' : j = 1, \ldots, n;\ k = 1, \ldots, m\}$$

and apply (17-1).

(17-3)$^+$ Let $\mathfrak{J}_o$ be the family of all sets of the form $\bigcup_{j=1}^{n} (F_j \times H_j)$, where $F_j \in \mathfrak{J}_1 \cap \mathfrak{G}_{1,\delta}$ and $H_j \in \mathfrak{J}_2 \cap \mathfrak{G}_{2,\delta}$, $j = 1, \ldots, n$. Show that

(i) $\mathfrak{J}_o$ is a sufficient family of compact sets in X.

(ii) If $C \in \mathfrak{J}_o$, then there are disjoint sets $A_k \times B_k \in \mathfrak{R}_o$, $k = 1, \ldots, m$, such that $C = \bigcup_{k=1}^{m} (A_k \times B_k)$. Hint: Use (16.1).

(17-4)$^+$ Let $\mathfrak{J}_o$ be the same family as in (17-3). If $C \in \mathfrak{J}_o$, then by (17-3)(ii), $C = \bigcup_{k=1}^{m} (A_k \times B_k)$, where $A_k \times B_k$, $k = 1, \ldots, m$, are disjoint sets from $\mathfrak{R}_o$. Let

$$v(C) = \sum_{k=1}^{m} \mu_1(A_k)\mu_2(B_k)$$

Without using (16.4), show that

(i) v is a well-defined function on $\mathfrak{J}_o$. Hint: Use (17-2).

(ii) $(X, \mathfrak{J}_o, v)$ is a volume space.

(17-5)$^+$ If $(X, \mathfrak{J}_o, v)$ is the volume space from (17-4), show that

(i) $\mathfrak{J}_o \subset \mathfrak{J}$ and $v(C) = \iota(C)$ for each $C \in \mathfrak{J}_o$. Hint: Use (16.4).

(ii) $\mathfrak{J}_o \subset \mathfrak{J} \cap \mathfrak{G}_\delta$ and $v(C) = \nu(C)$ for each $C \in \mathfrak{J}_o$. Hint: If $C \in \mathfrak{J}_o$ find functions $f_n \in C_o(X)$, $n = 1, 2, \ldots$, such that $f_n \searrow \chi_C$. Observe that $\int_X f_n \, d\nu = \int_X f_n \, d\iota$, and use (i) and (5.17).

(iii) If w is the inner volume induced by the volume v, then $w(G) = \nu(G)$ for each $G \in \mathfrak{G}$.

(iv) The measure space $(X, \mathfrak{N}, \nu)$ is induced by the volume v. Hint: Observe that if μ_v is the outer measure induced by v, then $\mu_v(G) = \nu(G)$ for each $G \in \mathfrak{G}$. Use this, (13-10)(iii), and (9.10).

(17-6)*+ In (17.11) we proved that $C_o(X)$ is a product system with respect to $(C_o(X_1), J_1)$ and $(C_o(X_2), J_2)$ by applying (16.6). Prove it again by using (17.7) instead of (16.6). Observe that this way the construction of the topological product of measures can be made independent of the results from Chapter 16.

Recall that counting, weighted counting, and Dirac measures are defined in (8-1).

(17-7)+ Let μ_1 and μ_2 be the Dirac measures at $x_o \in X_1$ and $y_o \in X_2$, respectively. Show that

(i) $\mathfrak{N} = \exp X$ and ν is the Dirac measure at (x_o, y_o). Hint: Use (17.8).

(ii) ι is the Dirac measure at (x_o, y_o) on $\mathfrak{I}$. Hint: Use (16-5)(ii) and (16-3)(iii).

(iii) If $\{x_o\} \in \mathfrak{M}_1$ and $\{y_o\} \in \mathfrak{M}_2$, then $\mathfrak{I} = \exp X$. Hint: Use the completeness of ι.

(17-8)+ Let X_i be a discrete space, and let μ_i be the weighted counting measure determined by a real-valued function f_i on X_i, $i = 1, 2$. Prove the following:

(i) $\mathfrak{N} = \exp X$ and ν is the weighted counting measure determined by $f_1 \otimes f_2$. Hint: Use (17.8).

(ii) $(X, \mathfrak{I}, \iota) = (X, \mathfrak{N}, \nu)$. Hint: Use (16-5)(ii), (16-3)(iii), and (8-11)(vi).

(17-9) Let $(X_1, \mathfrak{M}_1, \mu_1) = (\mathbf{R}, \Lambda, \lambda)$ [see (10-7)], and let $(X_2, \mathfrak{M}_2, \mu_2) = (Y, \exp Y, \varkappa)$, where Y is an uncountable discrete space and $\varkappa$ is the counting measure. Show that

(i) If $A \in \mathfrak{I}$ is such that $A^y \neq \emptyset$ for uncountably many $y \in X_2$, then $\iota(A) = +\infty$.

(ii) ι is a regular measure. Hint: Use (i) and the regularity of λ.

(iii) The measure spaces $(X, \mathfrak{J}, \iota)$ and $(X, \mathfrak{N}, \nu)$ coincide with the measure space $(X, \mathfrak{J}, \iota)$ from (11-3). Hint: Use (ii), (17.15) and (9.10).

(17-10) Let X_1 be an uncountable discrete space, $\mathfrak{M}_1 = \exp X_1$, and let μ_1 be the counting measure. Let X_2 be a singleton and $\mu_2 = 0$. Show that

(i) $\mathfrak{J} = \mathfrak{N} = \exp X$ and $\nu = 0$.

(ii) For $A \subset X$, $\iota(A) = 0$ if A is countable, and $\iota(A) = +\infty$ if A is uncountable. In particular, the measure ι is not regular.

In exercises (17-11) and (17-12) we shall assume that μ_1 and μ_2 are regular measures.

(17-11)$^+$ Prove that $\nu = 0$ iff $\mu_1 = 0$ or $\mu_2 = 0$. Hint: Use (17.8).

(17-12)$^{*+}$ Show that

(i) If ι is σ-finite, then so is ν. Hint: Use (17.15).

(ii) If μ_1 and μ_2 are σ-finite, then so is ν. Hint: Use (i) and (16-4)(i) and (ii).

(iii) If $\mu_i \neq 0$, $i = 1, 2$, and ν is σ-finite, then so are μ_1 and μ_2. Hint: By (8-9)(iv), we may assume that μ_1 and μ_2 are complete measures. If $\{A_n\} \subset \mathfrak{N}$, $A_n \nearrow X$, and $\nu(A_n) < +\infty$, $n = 1, 2, \ldots$, then by (17.14), there is an $M \in \mathfrak{M}$ such that $\mu_1(M) = 0$ and $(A_n)_x \in \mathfrak{M}$ for each $x \in X_1 - M$, $n = 1, 2, \ldots$. Observe that

$$X_1 - M = \bigcup_{n=1}^{\infty} \{x \in X_1 - M : \mu_2[(A_n)_x] > 0\}$$

whenever $\mu_2(X_2) > 0$. Use (17.14) and (10.8) to show that the sets $\{x \in X_1 - M : \mu_2[(A_n)_x] > 0\}$, $n = 1, 2, \ldots$, are μ_1-σ-finite.

(iv) If $\mu_i \neq 0$, $i = 1, 2$, then ν is σ-finite iff ι is σ-finite.

If Y_i is a set and $\mathfrak{S}_i \subset \exp Y_i$, $i = 1, 2$, we denote by $\mathfrak{S}_1 \times \mathfrak{S}_2$ the σ-algebra in $Y_1 \times Y_2$ generated by the family

$$\{C_1 \times C_2 : C_i \in \mathfrak{S}_i, \ i = 1, 2\}$$

It follows from (16-3)(iii), (16-4)(i) and (ii), and (8.16) that if μ_1 and μ_2 are σ-finite, then ι is the completion [see (8-9)] of its restriction to $\mathfrak{M}_1 \times \mathfrak{M}_2$. If μ_1 and μ_2 are σ-finite and regular, then by (17.16) the regularity of ι is determined by the size of $\mathfrak{J}$. Thus it is important to have some idea how large the system $\mathfrak{M}_1 \times \mathfrak{M}_2$ can be.

<u>(17-13)</u>$^+$ Let $\mathfrak{B}_i$ be the Borel σ-algebra in a topological space Y_i, $i = 1, 2$, and let $\mathfrak{B}$ be the Borel σ-algebra in $Y_1 \times Y_2$. Prove that

(i) $\mathfrak{B}_1 \times \mathfrak{B}_2 \subset \mathfrak{B}$. <u>Hint</u>: For $i = 1, 2$ the projections $\pi_i : Y_1 \times Y_2 \to Y_i$ [see (1-2)] are continuous, and consequently $(\mathfrak{B},\mathfrak{B}_i)$-measurable [see (7.7)]. If $B_1 \in \mathfrak{B}_1$ and $B_2 \in \mathfrak{B}_2$, observe that $B_1 \times B_2 = \pi_1^{-1}(B_1) \cap \pi_2^{-1}(B_2)$.

(ii) If Y_1 and Y_2 are second countable, then $\mathfrak{B}_1 \times \mathfrak{B}_2 = \mathfrak{B}$.

Exercises (17-16)(iii) and (17-17) show that $\mathfrak{B}_1 \times \mathfrak{B}_2 \neq \mathfrak{B}$ in general.

<u>(17-14)</u>$^+$ Prove that a locally compact Hausdorff space is second countable iff it is metrizable and σ-compact. Use this, (16.4), and (17-13)(ii) to give an alternate proof of (17.17).

<u>(17-15)</u>$^+$ Let Y be a set, and let $\Delta = \{(y,y) : y \in Y\}$ be the diagonal in $Y \times Y$. A family $\mathfrak{S} \subset \exp Y$ is called a <u>separating family</u> in Y if for each pair of distinct points $y, z \in Y$, there is a $C \in \mathfrak{S}$ such that $y \in C$ and $z \notin C$. Prove the following:

(i) If $\mathfrak{S} \subset \exp Y$ and $\Delta \in \mathfrak{S} \times \mathfrak{S}$, then $\mathfrak{S}$ is a separating family in Y.

(ii) If $\mathfrak{S} \subset \exp Y$ and $\Delta \in \mathfrak{S} \times \mathfrak{S}$, then there is a countable $\mathfrak{S}_o \subset \mathfrak{S}$ which is a separating family in Y. <u>Hint</u>: Use (i) and (7-7)(ii).

(iii) If $\mathfrak{S}$ is a separating family in Y, then $\|Y\| \leq \|\exp \mathfrak{S}\|$. <u>Hint</u>: Observe that the map $y \mapsto \{C \in \mathfrak{S} : y \in C\}$ from Y to $\exp \mathfrak{S}$ is one-to-one.

(iv) $\Delta \in (\exp Y) \times (\exp Y)$ iff $\|Y\| \leq \mathfrak{c}$. <u>Hint</u>: If $\|Y\| \leq \mathfrak{c}$, map Y injectively into $\mathbf{R}$ and use (17-13)(ii). If $\|Y\| > \mathfrak{c}$, apply (ii) and (iii).

<u>Note</u>: The reader should compare this exercise with Ref. 23.

<u>(17-16)</u>$^{*+}$ Let $\mathfrak{B}_{o,i}$ be the Baire σ-algebra [see (7-12)] in a locally compact Hausdorff space X_i, $i = 1, 2$, and let $\mathfrak{B}_o$ be the Baire σ-algebra in $X = X_1 \times X_2$. Prove that

(i) $\mathfrak{B}_{o,1} \times \mathfrak{B}_{o,2} \subset \mathfrak{B}_o$. Hint: Apply the hint from (17-13)(i).

(ii) If X_1 and X_2 are σ-compact, then $\mathfrak{B}_{o,1} \times \mathfrak{B}_{o,2} = \mathfrak{B}_o$. Hint: Use (7-13)(iii) and (17.1).

(iii) If X_i is a discrete space and $\|X_i\| > c$, $i = 1, 2$, then $\mathfrak{B}_{o,1} \times \mathfrak{B}_{o,2} \neq \mathfrak{B}_o$. Hint: Use (17-15)(iii).

(17-17) Let $X_1 = X_2 = W$ [see Chapter 1, Section B], let $\mathfrak{B}_i$ be the Borel σ-algebra in X_i , $i = 1, 2$, and let $\mathfrak{B}$ be the Borel σ-algebra in $X = X_1 \times X_2$. Show that $\mathfrak{B}_1 \times \mathfrak{B}_2 \neq \mathfrak{B}$. Hint: Let $\mu_1 = \mu_2 = \mu$ where μ is the measure from (9-10)(v), and consider the open set $G = \{(\alpha, \beta) \in X : \alpha < \beta\}$. Observe that $G \in \mathfrak{J}$ contradicts (16.8), and conclude from (16.4) that $G \notin \mathfrak{B}_1 \times \mathfrak{B}_2$.

18. REGULARITY OF BOREL MEASURES

The regular measures were introduced and discussed in Chapter 9. Their importance was demonstrated later on vis-à-vis the Riesz representation theorem [Theorem (13.9)]. In this chapter we shall present some topological and set-theoretic criteria under which a topological space admits a large family of regular measures.

Throughout we shall assume that X is a locally compact Hausdorff space. By $\mathfrak{J}$, $\mathfrak{G}$, and $\mathfrak{B}$ we shall denote, respectively, the families of all compact, open, and Borel subsets of X.

(18.1) Definition. Let μ be a measure defined on a σ-algebra $\mathfrak{M}$ in X. The measure $\mu^\sim$ which is the completion and saturation (in this order) of μ is called the natural extension of μ. The domain of $\mu^\sim$ is denoted by $\mathfrak{M}^\sim$.

Recall that the completion and saturation of a measure was defined in exercises (8-9) and (8-10), respectively. Using the notation of these exercises we can write

$$\mu^\sim = (\bar{\mu})^\wedge \qquad \text{and} \qquad \mathfrak{M}^\sim = (\bar{\mathfrak{M}})^\wedge$$

It follows from (8-9)(ii) and (8-10)(ii), (iii) that $\mu^\sim$ is a complete saturated measure. Moreover, if $\mathfrak{G} \subset \mathfrak{M}$, then by (9-5) and (9-6)(i), $\mu^\sim$ is regular iff μ is.

(18.2) Proposition. Let $\mathfrak{M}$ be a σ-algebra in X containing $\mathfrak{B}$, let μ be a measure on $\mathfrak{M}$, and let ν be the restriction of μ to $\mathfrak{B}$. Then μ is regular iff ν is regular, $\mathfrak{M} \subset \mathfrak{B}^\sim$, and $\mu(A) = \nu^\sim(A)$ for each $A \in \mathfrak{M}$.

Proof. If μ is regular, then ν is regular, and hence, so is $\nu^\sim$. Moreover, since

$$\mu(F) = \nu(F) = \nu^\sim(F)$$

for each $F \in \mathfrak{F}$, it follows from (9.9) that $\mathfrak{M} \subset \mathfrak{B}^{\sim}$ and $\mu(A) = \nu^{\sim}(A)$ for each $A \in \mathfrak{M}$.

If ν is regular, so is $\nu^{\sim}$. Thus if ν is regular, $\mathfrak{M} \subset \mathfrak{B}^{\sim}$, and $\mu(A) = \nu^{\sim}(A)$ for each $A \in \mathfrak{M}$, it follows that also μ is regular. ///

The previous proposition is important: it shows that in studying the regularity of measures we can restrict our attention only to measures defined on the Borel σ-algebra $\mathfrak{B}$. We recall that a measure μ on $\mathfrak{B}$ is called Borel whenever $\mu(F) < +\infty$ for each $F \in \mathfrak{F}$ [see (9-4)]. Clearly, each regular measure on $\mathfrak{B}$ is Borel, but the converse is generally false [see (9-10)(vi)]. Our task will be to give some sufficient conditions under which all Borel measures in X are regular.

(18.3) Definition. Let μ be a Borel measure in X. A set $A \in \mathfrak{B}$ is called μ-inner regular if

$$\mu(A) = \sup\{\mu(F) : F \in \mathfrak{F},\ F \subset A\}$$

and μ-outer regular if

$$\mu(A) = \inf\{\mu(G) : G \in \mathfrak{G},\ A \subset G\}$$

Thus a Borel measure μ in X is regular iff each open set is μ-inner regular and each Borel set is μ-outer regular. If μ is a σ-finite regular Borel measure, then by (9.3) each Borel set is both μ-inner and μ-outer regular. When no confusion is possible, we shall say only that a set $A \in \mathfrak{B}$ is inner or outer regular instead of μ-inner or μ-outer regular, respectively.

(18.4) Definition. A Borel measure μ in X is called inner regular whenever each open subset of X is μ-inner regular.

We shall see that the inner regularity of a Borel measure sometimes implies its regularity.

(18.5) Lemma. Let μ be a Borel measure in X, and let $\{A_n\} \subset \mathfrak{B}$ be a sequence of outer regular sets. Then $A = \bigcup_{n=1}^{\infty} A_n$ is outer regular.

Proof. If $\mu(A) = +\infty$, then A is outer regular. Hence, let $\mu(A) < +\infty$ and let $\epsilon > 0$. Then $\mu(A_n) < +\infty$ and we can find $G_n \in \mathfrak{G}$ such that $A_n \subset G_n$ and $\mu(G_n) < \mu(A_n) + \epsilon 2^{-n}$, $n = 1, 2, \ldots$. Setting $G = \bigcup_{n=1}^{\infty} G_n$

we have $G \in \mathfrak{G}$, $A \subset G$, and

$$\mu(G) - \mu(A) = \mu(G - A) = \mu(\bigcup_{n=1}^{\infty} G_n - \bigcup_{n=1}^{\infty} A_n) \leq \mu[\bigcup_{n=1}^{\infty} (G_n - A_n)]$$

$$\leq \sum_{n=1}^{\infty} \mu(G_n - A_n) \leq \epsilon \sum_{n=1}^{\infty} 2^{-n} = \epsilon \quad ///$$

(18.6) Lemma. Let μ be a Borel measure in X, and let $\{A_n\} \subset \mathfrak{B}$ be a decreasing sequence of outer regular sets. If $\mu(A_1) < +\infty$, then $A = \bigcap_{n=1}^{\infty} A_n$ is outer regular.

Proof. Choose $\epsilon > 0$. By (8.6) there is an integer $N \geq 1$ such that $\mu(A_N) < \mu(A) + \epsilon$. Since A_N is outer regular, we can find $G \in \mathfrak{G}$ such that $A \subset A_N \subset G$ and $\mu(G) < \mu(A) + \epsilon$. ///

(18.7) Lemma. Let μ be a Borel measure in X, let $A \in \mathfrak{B}$ be outer regular, and let $B \in \mathfrak{B}$ be inner regular. If $B \subset A$ and $\mu(A) < +\infty$, then $A - B$ is outer regular.

Proof. Given $\epsilon > 0$, there are $G \in \mathfrak{G}$ and $F \in \mathfrak{F}$ such that $A \subset G$, $F \subset B$, and

$$\mu(G) < \mu(A) + \frac{\epsilon}{2} \qquad \mu(F) > \mu(B) - \frac{\epsilon}{2}$$

Hence $G - F \in \mathfrak{G}$, $A - B \subset G - F$, and

$$\mu(G - F) = \mu(G) - \mu(F) < \mu(A) - \mu(B) + \epsilon = \mu(A - B) + \epsilon \quad ///$$

Recall that the definition of an algebra in X was given in exercise (7-4).

(18.8) Lemma. The collection $\mathfrak{A}$ of sets

$$\bigcup_{i=1}^{n} (G_i - H_i)$$

where $G_i, H_i \in \mathfrak{G}$ and $H_i \subset G_i$, $i = 1, \ldots, n$, is the algebra in X generated by $\mathfrak{G}$.

Proof. Since $\mathfrak{G} \subset \mathfrak{A}$ and $\mathfrak{A}$ is contained in the algebra generated by $\mathfrak{G}$, it suffices to show that $\mathfrak{A}$ is an algebra. Clearly, $\mathfrak{A}$ is closed with respect to the formation of finite unions and $X = X - \emptyset$ belongs to $\mathfrak{A}$. If $A = \bigcup_{i=1}^{n} (G_i - H_i)$, then

$$X - A = \bigcap_{i=1}^{n} [(X - G_i) \cup H_i]$$

This last expression is the union of the sets

$$[\bigcap_{j=1}^{k} (X - G_{i_j})] \cap [\bigcap_{j=k+1}^{n} H_{i_j}] = (X - \bigcup_{j=1}^{k} G_{i_j}) \cap [\bigcap_{j=k+1}^{n} H_{i_j}]$$

$$= \bigcap_{j=k+1}^{n} H_{i_j} - [\bigcup_{j=1}^{k} G_{i_j}] \cap [\bigcap_{j=k+1}^{n} H_{i_j}]$$

where k is an integer, $0 \le k \le n$, and $(i_1, i_2, \ldots, i_n)$ is a permutation of $(1, 2, \ldots, n)$. Thus $\mathfrak{A}$ is also closed with respect to the formation of complements. ///

With a bit more work one can show that in the previous lemma the sets $G_i - H_i$, $i = 1, \ldots, n$, may be assumed disjoint [see (7-5)]. However, we shall not need this stronger result.

(18.9) Proposition. Each finite inner regular Borel measure in X is regular.

Proof. Let μ be a finite inner regular Borel measure in X, and let $\mathfrak{O}$ be the family of all outer regular Borel sets. Clearly $\mathfrak{G} \subset \mathfrak{O}$. Thus by (18.8), (18.7), and (18.5), the algebra $\mathfrak{A}$ generated by $\mathfrak{G}$ is contained in $\mathfrak{O}$. It follows from (18.5), (18.6), and (7-6)(ii) that $\mathfrak{B} = \mathfrak{O}$. ///

To extend the previous proposition to Borel measures which are not finite is a nontrivial task. First we shall need some definitions.

Let $\mathcal{V} \subset \exp X$. If $x \in X$, then the set

$$st(x, \mathcal{V}) = \{V \in \mathcal{V} : x \in V\}$$

is called the star of x in $\mathcal{V}$. Thus $\mathcal{V}$ is a cover of X iff $st(x, \mathcal{V}) \neq \emptyset$ for each $x \in X$. We shall say that $\mathcal{V}$ is point-finite whenever $st(x, \mathcal{V})$ is finite for each $x \in X$.

Recall that a refinement of a family $\mathfrak{U}$ of sets was defined in (12.5).

(18.10) Definition. A topological space X is called metacompact if each open cover of X has an open point-finite refinement.

A locally finite cover of a topological space is clearly point-finite. Thus each paracompact space is metacompact. However, the converse is false [see (18-9)].

(18.11) Theorem. Let X be metacompact. Then each σ-finite inner regular Borel measure in X is regular.

Proof. (a) For each $x \in X$ choose an open neighborhood U_x of x so that U_x^- is compact. Because X is metacompact, the open cover $\{U_x : x \in X\}$ of X has an open point-finite refinement $\mathcal{V} = \{V_\alpha : \alpha \in T\}$. Thus if

$$B_n = \{x \in X : \| st(x,\mathcal{V}) \| = n\}$$

$n = 1, 2, \ldots$, then $\bigcup_{n=1}^{\infty} B_n = X$. Since the sets $\{x \in X : \| st(x,\mathcal{V}) \| \geq n\}$, $n = 1, 2, \ldots$, are open, the sets

$$B_n = \{x \in X : \| st(x,\mathcal{V}) \| \geq n\} - \{x \in X : \| st(x,\mathcal{V}) \| \geq n + 1\}$$

are Borel. Let $\mathfrak{D}_n = \{D \in \exp T : \|D\| = n\}$, and for $D \in \mathfrak{D}_n$, set

$$B_{n,D} = B_n \cap \bigcap\{V_\alpha : \alpha \in D\}$$

Clearly, the sets $B_{n,D}$ are disjoint, Borel, and

$$B_n = \bigcup\{B_{n,D} : D \in \mathfrak{D}_n\}$$

$n = 1, 2, \ldots.$

(b) Let μ be a σ-finite inner regular Borel measure in X. There are sets $X_k \in \mathfrak{B}$ such that $\mu(X_k) < +\infty$, $k = 1, 2, \ldots$, and $\bigcup_{k=1}^{\infty} X_k = X$. It follows from (5-8) that for each $n,k = 1, 2, \ldots$ the family

$$\mathfrak{D}_{n,k} = \{D \in \mathfrak{D}_n : \mu(B_{n,D} \cap X_k) > 0\}$$

is countable. Hence the set

$$T_o = \bigcup_{n,k=1}^{\infty} (\bigcup\{D : D \in \mathfrak{D}_{n,k}\})$$

is also countable. We shall show that $\mu(V_\alpha) = 0$ for each $\alpha \in T - T_o$.

(c) Let $\alpha \in T - T_o$, and let n and k be positive integers. By the definition of T_o, $\mu(B_{n,D} \cap X_k) = 0$ for each $D \in \mathfrak{D}_n$ for which $\alpha \in D$. Let F be a compact subset of $V_\alpha \cap B_n \cap X_k$. Because

$$V_\alpha \cap B_n = \bigcup\{B_{n,D} : D \in \mathfrak{D}_n, \alpha \in D\}$$

and because $B_{n,D}$ are open in B_n, the set F is covered by finitely many sets $B_{n,D} \cap X_k$ with $D \in \mathfrak{D}_n$ and $\alpha \in D$. Therefore, $\mu(F) = 0$. Since $V_\alpha \in \mathfrak{G}$ and $V_\alpha^- \in \mathfrak{F}$, the measure μ restricted to V_α is inner regular and finite. It follows from (18.9) and (9.2) that $\mu(V_\alpha \cap B_n \cap X_k) = 0$. Consequently, $\mu(V_\alpha) = 0$; for $V_\alpha = \bigcup_{n,k=1}^\infty (V_\alpha \cap B_n \cap X_k)$.

(d) Let $T_o = \{\alpha_1, \alpha_2, \ldots\}$, and set $G_n = V_{\alpha_n}$, n = 1, 2, ..., and $G_o = \bigcup\{V_\alpha : \alpha \in T - T_o\}$. Using (c) and the inner regularity of μ, we obtain that $\mu(G_o) = 0$. Thus for each n = 0, 1, ..., the measure μ restricted to G_n is finite and inner regular. By (18.9) each Borel subset of G_n is μ-outer regular. If $B \in \mathfrak{B}$, then $B = \bigcup_{n=0}^\infty (B \cap G_n)$, and the theorem follows from (18.5). ///

We note that the assumptions of the metacompactness and σ-finitene are essential in the previous theorem [see (18-3) and (18-1)].

<u>(18.12)</u> <u>Definition</u>. A Borel measure μ in X is called <u>strongly irregula</u> whenever there is an open set $G \subset X$ such that $\mu(G) = 1$ and $\mu(F) = 0$ for each compact set $F \subset G$.

Notice that the Borel measure ν from exercise (9-10)(vi) is strongly irregular.

The following lemma is quite useful.

<u>(18.13)</u> <u>Lemma</u>. Let μ be a σ-finite Borel measure in X which is not inner regular. Then there is a Borel measure ν in X which is strongly irregular.

<u>Proof</u>. By our assumption, there is a set $G \in \mathfrak{G}$ with $\mu(G) > \alpha$, where

$$\alpha = \sup\{\mu(F) : F \in \mathfrak{F}, F \subset G\}$$

We can find an increasing sequence $\{F_n\}$ of compact subsets of G suc that

$$\mu\left(\bigcup_{n=1}^{\infty} F_n\right) = \lim \mu(F_n) = \alpha$$

Let $C = \bigcup_{n=1}^{\infty} F_n$. Because μ is σ-finite, there is an increasing sequence $\{X_n\} \subset \mathfrak{B}$ such that $\mu(X_n) < +\infty$, $n = 1, 2, \ldots$, and $\bigcup_{n=1}^{\infty} X_n = X$. We have

$$\mu(G) = \mu\left[\bigcup_{n=1}^{\infty} (X_n \cap G)\right] = \lim \mu(X_n \cap G)$$

Therefore, $\mu(X_n \cap G) > \alpha$ for some integer $N \geq 1$. Let

$$\nu(B) = \frac{\mu[X_N \cap (B \cap G - C)]}{\mu[X_N \cap (G - C)]}$$

for each $B \in \mathfrak{B}$. Since

$$0 < \mu(X_N \cap G) - \mu(X_N \cap C) = \mu[X_N \cap (G - C)] < +\infty$$

ν is a well-defined Borel measure in X. Let $F \in \mathfrak{F}$ and $F \subset G$. Then

$$\mu\left(F \cup \bigcup_{n=1}^{k} F_n\right) \leq \alpha$$

for all $k = 1, 2, \ldots$. Hence,

$$\alpha + \mu(F - C) = \mu(C) + \mu(F - C) = \mu(F \cup C) = \lim_{k\to\infty} \mu\left(F \cup \bigcup_{n=1}^{k} F_n\right) \leq \alpha$$

and it follows that $\mu(F - C) = 0$. Therefore $\nu(F) = 0$ for each compact set $F \subset G$. Since $\nu(G) = 1$, ν is strongly irregular. ///

If the measure μ is not σ-finite, the previous lemma is generally false [see (18-4)].

Combining Lemma (18.13) with Proposition (18.9) and Theorem (18.11), we obtain the following corollary.

(18.14) Corollary. Let there be no strongly irregular Borel measure in X. Then each finite Borel measure in X is regular. If, in addition, X is metacompact, then each σ-finite Borel measure in X is regular.

This corollary provides a new proof for Corollary (13.16).

(18.15) Corollary. Let each open subset of X be σ-compact. Then all Borel measures in X are regular.

Proof. Because X is σ-compact, it is paracompact and each Borel measure in X is σ-finite. Since every open subset of X is σ-compact, there are no strongly irregular Borel measures in X. ///

In order to generalize Corollary (18.15) we shall introduce the following set-theoretic definition.

(18.16) Definition. A cardinal $\aleph$ is called measurable whenever there is a set T with $\|T\| = \aleph$ and a measure ν defined on exp T such that $\nu(T) = 1$ and $\nu(\{\alpha\}) = 0$ for each $\alpha \in T$.

The measure ν from the previous definition will be called a full probability in T. We shall make no attempt here to justify this terminology by explaining how the notion of a measure relates to our intuitive concept of probability. We shall merely refer the interested reader to the excellent exposition in Ref. 10, Section 44.

(18.17) Remark. We note that the usage of the term "measurable cardinal" may vary from author to author. For instance, the definition of a measurable cardinal given in Ref. 9, 12.1, p. 161 is different from ours. What we call a measurable cardinal is sometimes called a real-valued measurable cardinal (see, e.g., Ref. 21, Definition 1). The term "full probability" does not seem to be used in the literature.

A set $Y \subset X$ is called discrete if Y is a discrete space in the relative topology. Thus a set $Y \subset X$ is discrete iff for each $y \in Y$, there is a $G_y \in \mathfrak{G}$ such that $G_y \cap Y = \{y\}$.

The next proposition links the measurable cardinals with the regularity of Borel measures.

(18.18) Proposition. If X contains a discrete set of measurable cardinality, then there is a strongly irregular Borel measure in X.

Proof. Let $Y \subset X$ be a discrete set of measurable cardinality, and let ν be a full probability in Y. We define a Borel measure μ in X by setting $\mu(B) = \nu(B \cap Y)$ for each $B \in \mathfrak{B}$. Given $y \in Y$, choose $G_y \in \mathfrak{G}$ such that $G_y \cap Y = \{y\}$. Letting $G = \bigcup\{G_y : y \in Y\}$, we have $G \in \mathfrak{G}$ and $\mu(G) = \nu(Y) = 1$. Let $F \subset G$ be compact. Then F is covered by finitely many sets G_y, and so $F \cap Y$ is finite. Therefore, $\mu(F) = \nu(F \cap Y) = 0$. ///

Thus a necessary condition for all σ-finite Borel measures in X to be regular is that X contains no discrete set of measurable cardinality. In order to understand what this condition means we have to present some basic properties of measurable cardinals.

(18.19) Proposition. The countable cardinal $\aleph_0$ is not measurable. If a cardinal $\aleph$ is not measurable, then so is each cardinal $\aleph' \leq \aleph$.

Proof. It follows immediately from the σ-additivity that no full probability can be defined in a countable set.

If $A' \subset A$ and ν' is a full probability in A', we can define a full probability ν in A by letting $\nu(B) = \nu'(B \cap A')$ for each $B \subset A$. ///

(18.20) Lemma. Let $(Z, \mathfrak{M}, \mu)$ be a measure space, and let $\{A_\alpha : \alpha \in T\} \subset \mathfrak{M}$ be a disjoint collection satisfying the following conditions:

(i) $\bigcup\{A_\alpha : \alpha \in T'\} \in \mathfrak{M}$ for each $T' \subset T$

(ii) $\mu(A_\alpha) = 0$ for each $\alpha \in T$

(iii) $p = \mu(\bigcup\{A_\alpha : \alpha \in T\})$ is a positive real number.

Then T has a measurable cardinality.

Proof. Letting

$$\nu(T') = \frac{1}{p}\mu(\bigcup\{A_\alpha : \alpha \in T'\})$$

for each $T' \subset T$, we see immediately that ν is a full probability in T. ///

(18.21) Corollary. Let ν be a full probability in a set Z, $\{A_\alpha : \alpha \in T\} \subset \exp Z$, and let $\nu(A_\alpha) = 0$ for each $\alpha \in T$. If T has a nonmeasurable cardinality, then

$$\nu(\bigcup\{A_\alpha : \alpha \in T\}) = 0$$

Proof. Well order the set T, and let

$$B_\alpha = A_\alpha - \bigcup\{A_\beta : \beta \in T,\ \beta < \alpha\}$$

for each $\alpha \in T$. Then the collection $\{B_\alpha : \alpha \in T\}$ is disjoint and

$$\bigcup\{B_\alpha : \alpha \in T\} = \bigcup\{A_\alpha : \alpha \in T\}$$

Since $\nu(B_\alpha) = 0$ for each $\alpha \in T$, it follows from (18.20) that

$$\mu(\bigcup\{B_\alpha : \alpha \in T\}) = 0 \quad ///$$

(18.22) Proposition. Let T and A_α, $\alpha \in T$, be sets of nonmeasurable cardinalities. Then the cardinality of $A = \bigcup\{A_\alpha : \alpha \in T\}$ is also nonmeasurable.

Proof. Suppose that ν is a full probability in A. If $\nu(A_\alpha) > 0$ for some $\alpha \in T$, then $\nu/\nu(A_\alpha)$ restricted to the subsets of A_α is a full probability in A_α; a contradiction. If $\nu(A_\alpha) = 0$ for each $\alpha \in T$, we obtain a contradiction by applying (18.21). ///

If $\aleph$ is a cardinal we shall denote by $\aleph^+$ its immediate successor. Thus $\aleph^+$ is the least cardinal larger than $\aleph$.

(18.23) Theorem (Ulam). If $\aleph$ is a nonmeasurable cardinal, then so is $\aleph^+$.

Proof. Since all finite cardinals are nonmeasurable [see (18.19)], we shall assume that $\aleph$ is an infinite nonmeasurable cardinal. Let $\varkappa$ and $\varkappa^+$ be the first ordinals such that

$$\|\{\alpha : \alpha < \varkappa\}\| = \aleph \qquad \text{and} \qquad \|\{\alpha : \alpha < \varkappa^+\}\| = \aleph^+$$

Thus if $\alpha < \varkappa^+$, then

$$\|\{\beta : \beta < \alpha\}\| \le \aleph$$

and we can define a one-to-one map

$$f_\alpha : \{\beta : \beta < \alpha\} \to \{\gamma : \gamma < \varkappa\}$$

For $\xi < \varkappa$ and $\eta < \varkappa^+$, let

$$A(\xi,\eta) = \{\alpha < \varkappa^+ : \alpha > \eta,\ f_\alpha(\eta) = \xi\}$$

The sets $A(\xi,\eta)$ can be arranged into a transfinite matrix

$$\begin{array}{lllll} A(0,0) & A(0,1) & \cdots & A(0,\eta) & \cdots \\ A(1,0) & A(1,1) & \cdots & A(1,\eta) & \cdots \\ \cdots & \cdots & \cdots & \cdots & \cdots \\ A(\xi,0) & A(\xi,1) & \cdots & A(\xi,\eta) & \cdots \\ \cdots & \cdots & \cdots & \cdots & \cdots \end{array}$$

where the cardinality of the rows is $\aleph^+$ and the cardinality of the columns is $\aleph$. It is easy to see that the sets in each column are disjoint, and because the maps f_α are one-to-one also the sets in each row are disjoint. If $\eta < \varkappa^+$, let

$$A_\eta = \bigcup\{A(\xi,\eta) : \xi < \varkappa\}$$

and $B_\eta = \{\alpha < \varkappa^+ : \alpha \notin A_\eta\}$. Because $A_\eta = \{\alpha < \varkappa^+ : \alpha > \eta\}$, we have $\|B_\eta\| \leq \aleph$. Suppose that there is a full probability ν in the set $\{\alpha : \alpha < \varkappa^+\}$. By (18.21), $\nu(B_\eta) = 0$ for each $\eta < \varkappa^+$. Thus $\nu(A_\eta) = 1$ and again by (18.21), for each $\eta < \varkappa^+$, there is a $\xi_\eta < \varkappa$ such that $\nu[A(\xi_\eta,\eta)] > 0$. In particular,

$$\|\{A(\xi,\eta) : \nu[A(\xi,\eta)] > 0\}\| = \aleph^+$$

Since our transfinite matrix has only $\aleph$ rows, it follows that there is a $\xi_0 < \varkappa$ such that the ξ_0-row contains $\aleph^+$ sets $A(\xi,\eta)$ with $\nu[A(\xi,\eta)] > 0$. These sets are disjoint and the cardinal $\aleph^+$ is uncountable (remember that $\aleph$ is an infinite cardinal). An application of (5-8) gives

$$\nu(\{\alpha : \alpha < \varkappa^+\}) = +\infty$$

which is impossible. ///

Let $\aleph_\alpha$ be the <u>first</u> measurable cardinal. By (18.19) and (18.23), α is a limit ordinal and consequently,

$$\{\beta : \beta < \omega_\alpha\} = \bigcup_{\gamma<\alpha} \{\beta : \beta < \omega_\gamma\}$$

Since $\|\{\beta : \beta < \omega_\gamma\}\| = \aleph_\gamma < \aleph_\alpha$ for each $\gamma < \alpha$, it follows from (18.22) that

$$\|\{\gamma : \gamma < \alpha\}\| = \aleph_\alpha$$

and so $\alpha = \omega_\alpha$. This clearly indicates that $\aleph_\alpha$ is a very large cardinal: for example, $\aleph_\alpha \geq \aleph_{\omega_{\omega_{\cdot_{\cdot_{\cdot}}}}}$, where $\omega_{\omega_{\cdot_{\cdot_{\cdot}}}}$ is the limit of a countable sequence of ordinals

$$\omega, \omega_\omega, \omega_{\omega_\omega}, \ldots$$

This is, however, not enough. In view of (18.22) we see immediately that $\aleph_\alpha > \aleph_{\omega_{\omega_{\cdot_{\cdot_{\cdot}}}}}$, and using more refined techniques, it can be shown that $\aleph_\alpha$ is actually much larger than $\aleph_{\omega_{\omega_{\cdot_{\cdot_{\cdot}}}}}$ (see Ref. 21, Theorem 1).

Thus measurable cardinals are rather strange. If they exist, they are extremely large. However, we do not know whether to assume their existence is consistent (it is consistent, though, to assume that they do not exist). On the other hand, without the continuum hypothesis we have no way to estimate how large the continuum is. Hence it is conceivable that the continuum is a measurable cardinal. In fact, to assume that a measurable cardinal exists is equiconsistent with assuming that the continuum is a measurable cardinal. The rigorous study of these questions penetrates deeply into the foundations of mathematics. We refer the interested reader to Ref. 21.

In conclusion, we feel that by eliminating the spaces containing a discrete set of measurable cardinality, we are making a theoretical restriction which is quite negligible from the practical point of view of a working mathematician.

The final result of this chapter is based on an idea of R. Haydon (see Ref. 12, Proposition 3.2). We shall apply it to weakly θ-refinable spaces which were recently defined by H. R. Bennett and D. J. Lutzer (see Ref. 3).

(18.24) Definition. A topological space X is called weakly θ-refinable if for each open cover $\mathcal{U}$ of X we can find families $\mathcal{V}_n \subset \exp X$, $n = 1, 2, \ldots$, with the following properties:

(i) $\bigcup_{n=1}^{\infty} \mathcal{V}_n$ is an open refinement of $\mathcal{U}$

(ii) For each $x \in X$, there is an integer $n_x \geq 1$ such that $\mathrm{st}(x, \mathcal{V}_{n_x})$ is nonempty and finite.

As customary, we shall say that X is hereditarily weakly θ-refinable whenever each subspace of X is weakly θ-refinable.

We note that although the term "weakly θ-refinable" sounds a bit artificial, it is used by topologists and has an established place in the contemporary hierarchy of topological spaces.

(18.25) Remark. If in Definition (18.24) we add the requirement that each family $\mathcal{V}_n$, $n = 1, 2, \ldots$, covers X, we obtain the definition of a θ-refinable space. Each metacompact space is clearly θ-refinable but not vice versa, as exercise (18-2) shows. With the exception of this exercise, θ-refinable spaces will not be used. We shall deal only with weakly θ-refinable spaces.

(18.26) Proposition. Let $X = \bigcup_{n=1}^{\infty} X_n$, where each subspace X_n, $n = 1, 2, \ldots$, is weakly θ-refinable. Then X is weakly θ-refinable.

Proof. Let $\mathcal{U}$ be an open cover of X. Then

$$\mathcal{U}_n = \{U \cap X_n : U \in \mathcal{U}\}$$

is an open (in X_n) cover of X_n, $n = 1, 2, \ldots$. By our assumption we can find families $\mathcal{V}_{n,k} \subset \exp X_n$, $n,k = 1, 2, \ldots$, such that $\bigcup_{k=1}^{\infty} \mathcal{V}_{n,k}$ is an open (in X_n) refinement of $\mathcal{U}_n$ and to each $x \in X_n$, there is an integer $k_x \geq 1$ for which $\mathrm{st}(x, \mathcal{V}_{n,k_x})$ is nonempty and finite. If $V \in \mathcal{V}_{n,k}$, then V is open in X_n and $V \subset U$ for some $U \in \mathcal{U}$. Choose an open (in X) set $V' \subset U$ with $V' \cap X_n = V$, and let

$$\mathcal{V}'_{n,k} = \{V' : V \in \mathcal{V}_{n,k}\}$$

$n, k = 1, 2, \ldots$. Then $\bigcup_{n,k=1}^{\infty} \mathcal{V}_{n,k}$ is an open (in X) refinement of $\mathcal{U}$. If $x \in X$, then $x \in X_{n_x}$ for some integer $n_x \geq 1$. Hence there is an integer $k_x \geq 1$ such that $st(x, \mathcal{V}_{n_x, k_x})$ is nonempty and finite. Since

$$\| st(x, \mathcal{V}'_{n_x, k}) \| = \| st(x, \mathcal{V}_{n_x, k}) \|$$

for each $k = 1, 2, \ldots$, the proposition follows. ///

A weakly θ-refinable space may not be hereditarily weakly θ-refinable. Indeed, take a space which is not weakly θ-refinable [see, e.g., (18-16)]; its one-point compactification is weakly θ-refinable but not hereditarily weakly θ-refinable.

(18.27) Proposition. Let X be weakly θ-refinable, and let Y be a closed subset of X. Then Y is weakly θ-refinable.

Proof. Let $\mathcal{U}$ be an open (in Y) cover of Y. For each $U \in \mathcal{U}$, choose an open (in X) set $U' \subset X$ such that $U' \cap Y = U$ and let

$$\mathcal{U}' = \{X - Y, U' : U \in \mathcal{U}\}$$

Since Y is closed in X, $\mathcal{U}'$ is an open (in X) cover of X. Thus we can find families $\mathcal{V}'_n \subset \exp X$, $n = 1, 2, \ldots$, such that $\bigcup_{n=1}^{\infty} \mathcal{V}'_n$ is an open (in X) refinement of $\mathcal{U}'$ and for each $x \in X$ there is an integer $n_x \geq 1$ for which $st(x, \mathcal{V}'_{n_x})$ is nonempty and finite. Let

$$\mathcal{V}_n = \{V' \cap Y : V' \in \mathcal{V}'_n\}$$

$n = 1, 2, \ldots$. Then $\bigcup_{n=1}^{\infty} \mathcal{V}_n$ is an open (in Y) refinement of $\mathcal{U}$. Because

$$\| st(y, \mathcal{V}_n) \| = \| st(y, \mathcal{V}'_n) \|$$

for each $y \in Y$ and $n = 1, 2, \ldots$, the proposition follows. ///

(18.28) Corollary. Let X be weakly θ-refinable, and let Y be an F_σ subset of X. Then Y is weakly θ-refinable.

This corollary follows immediately from (18.27) and (18.26).

The next proposition gives a useful criterion for a space to be hereditarily weakly θ-refinable.

(18.29) Proposition. Let each open subspace of X be weakly θ-refinable. Then X is hereditarily weakly θ-refinable.

Proof. Let $Y \subset X$ and let $\mathfrak{U}$ be an open (in Y) cover of Y. For each $U \in \mathfrak{U}$, choose an open (in X) set $U' \subset X$ such that $U' \cap X = U$. Let

$$\mathfrak{U}' = \{U' : U \in \mathfrak{U}\} \qquad \text{and} \qquad G = \bigcup\{U' : U' \in \mathfrak{U}'\}$$

Since G is open in X, we can find families $\mathcal{V}'_n \subset \exp G$, $n = 1, 2, \ldots,$ such that $\bigcup_{n=1}^{\infty} \mathcal{V}'_n$ is an open (in G, and hence in X) refinement of $\mathfrak{U}'$ and to each $x \in G$ there is an integer $n_x \geq 1$ for which $\mathrm{st}(x, \mathcal{V}'_{n_x})$ is nonempty and finite. Let

$$\mathcal{V}_n = \{V' \cap Y : V' \in \mathcal{V}'_n\}$$

$n = 1, 2, \ldots$. Then $\bigcup_{n=1}^{\infty} \mathcal{V}_n$ is an open (in Y) refinement of $\mathfrak{U}$. Because

$$\|\mathrm{st}(y, \mathcal{V}_n)\| = \|\mathrm{st}(y, \mathcal{V}'_n)\|$$

for each $y \in Y$ and $n = 1, 2, \ldots,$ the proposition follows. ///

The following technical lemma of Bennett and Lutzer has frequent applications.

(18.30) Lemma. A topological space X is weakly θ-refinable iff for each open cover $\mathfrak{U}$ of X we can find families $\mathcal{V}_n \subset \exp X$, $n = 1, 2, \ldots,$ with the following properties:

(i) $\bigcup_{n=1}^{\infty} \mathcal{V}_n$ is an open refinement of $\mathfrak{U}$

(ii) For each $x \in X$, there is an integer $n_x \geq 1$ such that $\|\mathrm{st}(x, \mathcal{V}_{n_x})\| = 1$.

Proof. Suppose that X is weakly θ-refinable, and let $\mathfrak{U}$ be an open cover of X. We can find families $\mathcal{V}_n \subset \exp X$, $n = 1, 2, \ldots,$ such that $\bigcup_{n=1}^{\infty} \mathcal{V}_n$ is an open refinement of $\mathfrak{U}$ and to each $x \in X$ there is an integer $n_x \geq 1$ for which $\mathrm{st}(x, \mathcal{V}_n)$ is nonempty and finite. For $n, k = 1, 2, \ldots,$ let $\mathcal{V}_{n,k}$ consist of all sets $V_1 \cap \ldots \cap V_k$ where $V_1, \ldots, V_k$ are distinct elements of $\mathcal{V}_n$. Clearly, $\bigcup_{n,k=1}^{\infty} \mathcal{V}_{n,k}$ is an open refinement of of $\mathfrak{U}$. Given $x \in X$, choose an integer $n \geq 1$ so that $\operatorname{st}(x, \mathcal{V}_n)$ is nonempty and finite. If $\|\operatorname{st}(x, \mathcal{V}_n)\| = k$, then $\|\operatorname{st}(x, \mathcal{V}_{n,k})\| = 1$.

The converse is obvious. ///

We are ready to prove the main theorem of this chapter.

(18.31) Theorem. Let X be a locally compact Hausdorff space which is hereditarily weakly θ-refinable and which contains no discrete subset of measurable cardinality. Then each finite Borel measure in X is regular. If, in addition, X is metacompact, then each σ-finite Borel measure in X is regular.

Proof. In view of (18.14), it suffices to show that there is no strongly irregular Borel measure in X. Hence, suppose that there is a Borel measure μ in X and a $G \in \mathfrak{G}$ such that $\mu(G) = 1$ and $\mu(F) = 0$ for each compact set $F \subset G$. For each $x \in G$, choose an open neighborhood U_x of x so that U_x^- is a compact subset of G. Because G is weakly θ-refinable, by (18.30) we can find families $\mathcal{V}_n \subset \exp G$, $n = 1, 2, \ldots,$ such that $\bigcup_{n=1}^{\infty} \mathcal{V}_n$ is an open (in G, and hence in X) refinement of $\{U_x : x \in G\}$, and to each $x \in G$ there is an integer $n_x \geq 1$ for which $\|\operatorname{st}(x, \mathcal{V}_{n_x})\| = 1$. Thus if

$$A_n = \{x \in G : \|\operatorname{st}(x, \mathcal{V}_n)\| = 1\}$$

$n = 1, 2, \ldots,$ then $G = \bigcup_{n=1}^{\infty} A_n$. Since $\mathcal{V}_n \subset \mathfrak{G}$, $n = 1, 2, \ldots,$ the sets

$$G_{n,k} = \{x \in G : \|\operatorname{st}(x, \mathcal{V}_n)\| \geq k\}$$

$k = 1, 2,$ are open. Hence the sets $A_n = G_{n,1} - G_{n,2}$, $n = 1, 2, \ldots,$ are Borel. By (8.9), $\mu(A_N) > 0$ for some integer $N \geq 1$. Clearly,

$$\mathfrak{W} = \{A_N \cap V \neq \emptyset : V \in \mathcal{V}_N\}$$

is a disjoint open (in A_N) cover of A_N. Thus any union of sets from $\mathfrak{W}$ is open in A_N, and therefore it is a Borel subset of X. Moreover, for each $W \in \mathfrak{W}$, W^- is a compact subset of G, and so $\mu(W) = 0$. By

(18.20), $\mathfrak{w}$ has a measurable cardinality. For each $W \in \mathfrak{w}$, choose a $y_W \in W$ and let $Y = \{y_W : W \in \mathfrak{w}\}$. If $W = A_N \cap V$ is from $\mathfrak{w}$, then

$$Y \cap V = (Y \cap A_N) \cap V = Y \cap W = \{y_W\}$$

It follows that Y is a discrete subset of X. This is a contradiction, for $\|Y\| = \|\mathfrak{w}\|$. ///

A locally compact Hausdorff space in which each finite Borel measure is regular is called a Radon space. Using this name, we can reformulate the first part of the previous theorem.

(18.32) Theorem. Each hereditarily weakly θ-refinable, locally compact, Hausdorff space which contains no discrete subset of measurable cardinality is a Radon space.

This theorem was proved independently by R. J. Gardner [see The regularity of Borel measures and Borel measure-compactness, Proc. London Math. Soc., (3)30(1975), 95-113], and it seems to be the most general result presently available. A complete characterization of Radon spaces in topological terms is an important open problem.

Note: The regularity of locally finite measures [see (9-12)] has been extensively studied in topological spaces which are not locally compact. However, the absence of local compactness requires techniques quite different from those employed in this chapter. We refer the interested reader to L. Schwartz, Radon Measures, Oxford University Press, London 1973.

Exercises

(18-1) Let Y be an uncountable discrete space, and let $X = \mathbf{R} \times Y$. If $B \subset X$ and $y \in Y$, let $B^y = \{x \in \mathbf{R} : (x,y) \in B\}$. Denote by λ the Lebesgue measure in $\mathbf{R}$ [see (10-7)] and set

$$\mu(B) = \Sigma\{\lambda(B^y) : y \in Y\}$$

for each Borel set $B \subset X$ [see (5-9)]. Show that

(i) μ is an inner regular Borel measure in X.

(ii) μ is not σ-finite. Hint: Use (5-9)(ii).

(iii) μ is not regular. Hint: Observe that $\mu(G) = +\infty$ for each open set $G \subset X$ with $\{0\} \times Y \subset G$.

(18-2)* Let $X \subset R^2$ consist of the points $(s,0)$ where $s \in [0,1]$, and $(k2^{-n}, 2^{-n})$, where $k = 0, 1, \ldots, 2^n$, $n = 0, 1, \ldots$. Show that we can define a topology in X as follows: The points $(k2^{-n}, 2^{-n})$ are isolated and a neighborhood base at $(s,0)$ is given by the sets

$$U(s,\varepsilon) = \{(u,v) \in X : 2|u - s| < v < \varepsilon\}$$

where $\varepsilon > 0$ (draw a picture). Give X this topology and prove

(i) X is a locally compact locally metrizable Hausdorff space.

(ii) X is not metrizable. Hint: Observe that X is separable but that it contains a nonseparable subspace.

(iii) If $\mathcal{U} = \{U(s,\varepsilon_s) : s \in [0,1]\}$, then there is an $x \in X$ such that $\|st(x,\mathcal{U})\| \geq \aleph_0$. Hint: Use the Baire category theorem (see, e.g., Ref. 8, Chapter XI, Section 10) to show that there is an $\varepsilon > 0$ such that the set $\{s \in [0,1] : \varepsilon_s > \varepsilon\}^-$ contains an open interval.

(iv) X is not metacompact. Hint: Use (iii).

(v) X is θ-refinable [see (18.25)]. Hint: If $\varepsilon : [0,1] \to (0,+\infty)$, let $\mathcal{U}(\varepsilon)$ consist of all singletons $\{x\}$ with x isolated in X and of all sets $U(s,\varepsilon(s))$ with $s \in [0,1]$. Observe that each open cover of X has a refinement $\mathcal{U}(\varepsilon)$ for some $\varepsilon : [0,1] \to (0,+\infty)$, and let $\mathcal{V}_n = \mathcal{U}(\varepsilon/n)$, $n = 1, 2, \ldots$.

(18-3)* Let X be the topological space from the previous exercise. Denote by $\mu_{k,n}$ the Dirac measure [see (8-1)(v)] at $(k2^{-n}, 2^{-n})$, $k = 0, 1, \ldots, 2^n$, $n = 0, 1, \ldots$, and set

$$\mu(B) = \sum_{n=0}^{\infty} 2^{-n} \sum_{k=0}^{2^n} \mu_{k,n}(B)$$

for each Borel set $B \subset X$. Show that

(i) μ is a σ-finite Borel measure in X. Hint: Use (8-12)(i) and observe that for every $s \in [0,1]$, the set $U(s,1)$ contains at most one point on each horizontal line.

(ii) μ is inner regular.

(iii) μ is not regular. Hint: Apply the hint from (18-2)(iii) to show that $\mu(G) = +\infty$ for each open set G containing $\{(s,0) : s \in [0,1]\}$.

(iv) supp $\mu = X$ [see (9-8)].

(18-4) Let X be a discrete space with $\|X\| = \aleph_1$. For $A \subset X$, let $\mu(A) = 0$ if A is countable and $\mu(A) = +\infty$ otherwise. Show that

(i) μ is a Borel measure in X which is neither σ-finite nor inner regular.

(ii) There is no extremely irregular Borel measure in X. Hint: Use (18.19) and (18.23).

Notice that formally the definitions of the regularity, inner regularity, outer regularity, strong irregularity, and support [see (9-8)] of a Borel measure have meaning in an arbitrary Hausdorff space which need not be locally compact. The Hausdorff separation property is needed for compact sets to be Borel.

(18-5)$^+$ Let X be a Hausdorff space. Prove

(i) If μ is an inner regular Borel measure in X, then supp μ exists [see (9-8)]. Hint: Observe that the union of all open μ-null subsets of X is a μ-null set.

(ii) Let X be locally compact. Each Borel measure in X has a support iff each Borel measure in X is inner regular. Hint: Let μ be a Borel measure in X such that

$$\mu(G) > \sup\{\mu(F) : F \in \mathfrak{J},\ F \subset G\}$$

for some open set $G \subset X$. Find a σ-compact set $C \subset G$ for which

$$\mu(C) = \sup\{\mu(F) : F \in \mathfrak{J},\ F \subset G\}$$

If $B \subset X$ is a Borel set, set $\nu(B) = \mu[B \cap (G - C)]$ and show that ν has no support.

(iii) Let X be metacompact locally compact, and let μ be an inner regular Borel measure in X. Then μ is σ-finite iff supp μ is σ-compact. Hint: Analyze the proof of Theorem (18.11).

Note that (i) and (iii) generalize the results from (9-8)(ii) and (iv), respectively.

(18-6) Let X be the space from exercise (18-2). Denote by $\mu_{n,k}$ the Dirac measure [see (8-1)(v)] at $(k2^{-n}, 2^{-n})$, $k = 0, 1, \ldots, 2^n$, $n = 0, 1, \ldots$. For a Borel set $B \subset X$, set

$$\nu(B) = \sum_{n=0}^{\infty} 2^{-2n} \sum_{k=0}^{2^n} \mu_{n,k}(B)$$

if B is countable, and $\nu(B) = +\infty$ otherwise. Show that

(i) ν is a Borel measure in X.

(ii) X is not ν-inner regular.

(iii) supp $\nu = X$ [see (9-8)].

Find a Borel measure in X which has no support.

(18-7) Let X be a Hausdorff space, and let $X = \bigcup_{n=1}^{\infty} X_n$, where X_n, $n = 1, 2, \ldots$, are Borel subsets of X. Suppose that in no X_n is there a strongly irregular Borel measure. Show that there is no strongly irregular Borel measure in X.

(18-8)* Suppose that μ and μ_n, $n = 1, 2, \ldots$, are Borel measures in a Hausdorff space X and that $\mu = \sum_{n=1}^{\infty} \mu_n$. Prove the following:

(i) Let A be a Borel subset of X which is μ_n-inner regular for each $n = 1, 2, \ldots$. Then A is μ-inner regular.

(ii) Let A be a Borel subset of X which is μ_n-outer regular for each $n = 1, 2, \ldots$. Then A is μ-outer regular or $\mu(G) = +\infty$ for each open set G containing A.

(iii) Let μ_n be σ-finite and regular for each $n = 1, 2, \ldots$. If X is locally compact and metacompact, then μ is σ-finite and regular. Hint: By (i), μ is inner regular. Use (18-5)(iii) to show that μ is σ-finite and apply (18.11).

(18-9) Let

$$X = \{(u,v) \in \mathbf{R}^2 : u \geq 0,\ v \geq 0,\ u + v > 0\}$$

Show that

(i) We can define a topology in X as follows: The points $(u,v) \in X$ with $u > 0$ and $v > 0$ are isolated, and neighborhood bases at the points $(u,0) \in X$ and $(0,v) \in X$ are given, respectively, by the sets

$$\{(u,t) : t \in [0,+\infty) - F\} \qquad \text{and} \qquad \{(t,v) : t \in [0,+\infty) - F\}$$

where F is a finite subset of $(0,+\infty)$.

(ii) With the topology from (i), X is a locally compact metacompact Hausdorff space which is not paracompact.

(18-10) Show that a discrete subset A of a topological space X is a Borel subset of X. Hint: Observe that A is closed in a suitably chosen open subset of X.

(18-11) Let X be a topological space in which all open subsets are σ-compact. Show that each discrete subset of X is countable.

(18-12)$^+$ Let ν be a full probability in a set Z, and let $\mathfrak{S} \subset \exp Z$ be a disjoint family of nonmeasurable cardinality. Show that

$$\nu(\cup\{C : C \in \mathfrak{S}\}) = \Sigma\{\nu(C) : C \in \mathfrak{S}\}$$

Hint: Use (5-8), (18.21), and the σ-additivity of ν.

(18-13)* Let X be a locally compact Hausdorff space which is hereditarily paracompact and contains no discrete set of measurable cardinality. Without using Theorems (18.11) and (18.31), show that each σ-finite Borel measure in X is regular. Hint: Use the fact that X is a free union of σ-compact spaces (see Ref. 8, Chapter XI, Theorem 7.3, p. 241).

The next three exercises will be concerned with the space W of all countable ordinals (see Chapter 1, Section B).

(18-14)$^+$ Show that there is a subset of W which is not Borel. Hint: Use (9-10)(iv) and (v), (18.19), and (18.23).

(18-15) For a Borel set $B \subset W$, let $\mu(B) = 0$ if B is countable and $\mu(B) = +\infty$ otherwise. Show that μ is a nonsaturated Borel measure in W. Hint: Use (18-14).

(18-16) Show that W contains an open subset which is not weakly θ-refinable. Hint: Use (9-10)(vi), (18.29), and (18.31).

Note: It follows from Ref. 8, Chapter VIII, Section 2, exercise 3, p. 163 and Ref. 3, Theorem 11 that, in fact, W is not weakly θ-refinable.

(18-17) Let $\aleph_\alpha$ be a nonmeasurable cardinal. Show that so is $\aleph_{\omega_\alpha}$.

Hint: Use (18.22) and (18.23) to prove by transfinite induction that $\aleph_\beta$ is a nonmeasurable cardinal for each $\beta < \omega_\alpha$.

(18-18)*+ Let $(X, \mathfrak{M}, \mu)$ be a measure space, let $\mathfrak{C} \subset \mathfrak{M}$, and let $\mathfrak{C}_+ = \{A \in \mathfrak{C} : \mu(A) > 0\}$. Prove the following:

(i) Let $\|st(x, \mathfrak{C})\| \leq n$ for a fixed integer $n \geq 1$ and each $x \in X$. If μ is finite, then $\mathfrak{C}_+$ is countable. Hint: Observe that $\Sigma\{\chi_A : A \in \mathfrak{C}\} \leq n$. Thus by (6-14)(i), $\Sigma\{\mu(A) : A \in \mathfrak{C}\} \leq n\mu(X)$. Apply (5-9)(ii).

(ii) Let $\mathfrak{C}$ be a point-finite family. If μ is finite, then $\mathfrak{C}_+$ is countable. Hint: If

$$X_n = \{x \in X : \|st(x, \mathfrak{C})\| \leq n\}$$

$n = 1, 2, \ldots$, then $\bigcup_{n=1}^{\infty} X_n = X$. Denote by μ_e the outer measure induced by μ [see (12-9)(ii)], and use (i) and (14-12)(ii) to show that the families

$$\mathfrak{C}_n = \{A \in \mathfrak{C} : \mu_e(A \cap X_n) > 0\}$$

$n = 1, 2, \ldots$, are countable. Observe that $\mathfrak{C}_+ = \bigcup_{n=1}^{\infty} \mathfrak{C}_n$.

(iii) Statement (ii) remains correct if μ is σ-finite. Hint: Use (ii) and (8-12)(ii).

A family $\mathfrak{C} \subset \exp X$ is called point-countable if $\|st(x, \mathfrak{C})\| \leq \aleph_0$ for each $x \in X$. Show that statement (ii) is generally false for a point-countable family $\mathfrak{C}$. Hint: Consider the measure space $(X, \mathfrak{M}, \mu)$ from (9-10) and the family $\mathfrak{C} = \{[\alpha, \Omega) : \alpha < \Omega\}$.

The reader should compare this exercise with exercise (8-16).

(18-19)+ A cover $\mathcal{V}$ of a set X is called minimal if $\mathcal{V}$ contains no proper subcover. Show that

(i) Each point-finite cover $\mathcal{V}$ of a set X has a minimal subcover.

Hint: Apply Zorn's lemma to the family of all subcovers of $\mathcal{V}$.

(ii) The point-finiteness of $\mathcal{V}$ is essential in (i). Hint: Let $X = \mathbf{R}$ and let

$$\mathcal{V} = \{[-n,n] : n = 1, 2, \ldots\}$$

In exercises (18-20) through (18-23) we shall assume that X is a locally compact Hausdorff space.

(18-20)* Let μ be a diffused strongly regular measure in X [see (8-11) and (9-14)]. Show that if X is metacompact, then μ is σ-finite. Hint: With no loss of generality we may assume that supp $\mu = X$. Use (18-19)(i) to find a minimal, open, point-finite cover $\mathfrak{U}$ of X such that $\mu(U) < +\infty$ for each $U \in \mathfrak{U}$. For each $U \in \mathfrak{U}$, choose

$$x_U \in U - \bigcup\{V \in \mathfrak{U} : V \neq U\}$$

and let $A = \{x_U : U \in \mathfrak{U}\}$. Observe that A is closed and discrete, and that $\mu(A) = 0$. Choose an open set G so that $A \subset G$ and $\mu(G) < +\infty$. Apply (18-18)(ii) to show that $\{G \cap U : U \in \mathfrak{U}\}$ is countable.

Note: Using more refined techniques, it can be shown that a diffused strongly regular measure in a space X is σ-finite whenever X is weakly θ-refinable (see W. F. Pfeffer, Some remarks on generalized Borel measures in topological spaces, Proceedings of 1977 Spring Topology Conference in Baton Rouge). Whether the weak θ-refinability of X is essential seems unknown.

(18-21) A Borel measure μ in X is called moderated if there are open sets $G_n \subset X$ such that $\mu(G_n) < +\infty$, $n = 1, 2, \ldots$, and $X = \bigcup_{n=1}^{\infty} G_n$. Prove the following:

(i) Let μ be a σ-finite Borel measure in X. If μ is outer regular, then it is moderated.

(ii) Let μ be an inner regular Borel measure in X. If μ is moderated, then it is regular. Hint: Follow part (d) of the proof of Theorem (18.11).

(iii) The measure μ from (18-3) is σ-finite, inner regular, but not moderated.

Use (ii) and (18-18)(iii) to give an alternate proof of (18.11).

(18-22) Let Y be a discrete space of measurable cardinality, let Z be the one-point compactification of the discrete space of positive integers, and let $X = Y \times Z$. For each $B \subset X$, set

$$\mu(B) = \sum_{n=1}^{\infty} \nu(\{y \in Y : (y,n) \in B\})$$

where ν is a full probability in Y. Prove the following:

(i) X is a locally compact metrizable space.

(ii) μ is a σ-finite Borel measure in X.

(iii) μ is not moderated. Hint: Let $G_1, G_2, \ldots$ be open subsets of X such that $X = \bigcup_{n=1}^{\infty} G_n$. If n and k are positive integers, let

$$Y_{n,k} = \{y \in Y : \{y\} \times \{n,n+1,\ldots\} \subset G_k\}$$

Observe that $Y = \bigcup_{n,k=1}^{\infty} Y_{n,k}$, and conclude from this that $\mu(G_k) = +\infty$ for some integer $k \geq 1$.

(18-23)* Suppose that X contains no discrete subset of measurable cardinality, and that μ is a σ-finite Borel measure in X. Let $\mathfrak{U}$ be a point-finite family of open μ-null subsets of X [see (8.7)]. Prove that $G = \bigcup\{U : U \in \mathfrak{U}\}$ is μ-null. Hint: Assume that μ is finite and that $\mu(G) > 0$. For $n = 1, 2, \ldots$, let

$$G_n = \{x \in G : \|st(x,\mathfrak{U})\| = n\}$$

Observe that G_n are Borel sets, and that $\mu(G_N) > 0$ for some integer $N \geq 1$. Let $\mathfrak{U}_N$ be the family of all nonempty sets $G_N \cap U_1 \cap \ldots \cap U_N$ where $U_1, \ldots, U_n$ are distinct elements of $\mathfrak{U}$. Use (18.20) to obtain a contradiction. If μ is σ-finite, apply (8-12)(ii).

(18-24)+ Let X be metacompact, and suppose that it contains no discrete subset of measurable cardinality. Show that each σ-finite Borel measure in X is moderated. Hint: Use (18-18)(iii) and (18-23).

Note: The hypotheses of the previous exercise are essential. By (18-3), the metacompactness of X cannot be replaced by θ-refinability, and by (18-22), the cardinality restriction cannot be omitted.

If in a set T there is a <u>two-valued full probability</u>, i.e., a full probability which takes only values 0 and 1, then $\|T\|$ is called a <u>two-valued measurable cardinal</u>. Clearly, each two-valued measurable cardinal is a measurable cardinal. Through the following five exercises we shall see that a partial converse to this is also true.

<u>(18-25)</u> Let ν be a two-valued full probability in a set T, and let $\mathfrak{D} \subset \exp T$ have a cardinality which is not two-valued measurable. Prove that

(i) If $\nu(A) = 0$ for each $A \in \mathfrak{D}$, then $\nu(\bigcup\{A : A \in \mathfrak{D}\}) = 0$. <u>Hint</u>: Follow the proof of (18.21).

(ii) If $\nu(A) = 1$ for each $A \in \mathfrak{D}$, then $\nu(\bigcap\{A : A \in \mathfrak{D}\}) = 1$. <u>Hint</u>: Use (i).

<u>(18-26)</u>* Let T be a set. For each $\alpha \in T$, let

$$\mathfrak{S}_\alpha = \{A \subset T : \alpha \in A\} \qquad \mathfrak{D}_\alpha = \{A \subset T : \alpha \notin A\}$$

and show that

(i) For each $S \subset T$,

$$\{S\} = (\bigcap\{\mathfrak{S}_\alpha : \alpha \in S\}) \cap (\bigcap\{\mathfrak{D}_\alpha : \alpha \in T - S\})$$

(ii) If $\|T\|$ is not a two-valued measurable cardinal, then neither is $\|\exp T\|$. <u>Hint</u>: Suppose that ν is a two-valued full probability in exp T, and let $S = \{\alpha \in T : \nu(\mathfrak{S}_\alpha) = 1\}$. Use (i) and (18-25)(ii) to show that $\|T\|$ is a two-valued measurable cardinal.

<u>(18-27)</u> It follows from (18.19) and (18-25)(ii) that c is not a two-valued measurable cardinal. Prove this directly by showing that there is no two-valued full probability in the interval [0,1].

<u>(18-28)</u>* Let $(Z, \mathfrak{M}, \mu)$ be a measure space. A set $A \in \mathfrak{M}$ is called an <u>atom</u> of μ if $\mu(A) > 0$ and for each $B \in \mathfrak{M}$ with $B \subset A$ either $\mu(B) = \mu(A)$ or $\mu(B) = 0$. Show that

(i) The measure μ has no atoms iff for each $A \in \mathfrak{M}$ the set $\{\mu(B) : B \in \mathfrak{M},\ B \subset A\}$ is equal to the interval $[0, a]$ where $a = \mu(A)$.

Hint: Suppose that there is an $A \in \mathfrak{M}$ and $\alpha \in (0,a)$ such that $\mu(B) \neq \alpha$ for each measurable set $B \subset A$. Let

$$\mathfrak{M}_1 = \{B \in \mathfrak{M} : B \subset A,\ \mu(B) < \alpha\}$$

and for $n = 1, 2, \ldots$, construct inductively families $\mathfrak{M}_n$ and sets $B_n \in \mathfrak{M}_n$ such that

$$\mu(B_n) > \sup\{\mu(B) : B \in \mathfrak{M}_n\} - \frac{1}{n}$$

and $\mathfrak{M}_{n+1} = \{B \in \mathfrak{M}_n : B_n \subset B\}$. Set $B_0 = \bigcup_{n=1}^{\infty} B_n$, and observe that $\mu(B) = \mu(B_0)$ for each $B \in \mathfrak{M}_1$ for which $B_0 \subset B$. Similarly, construct a set $C_0 \in \mathfrak{M}$ such that $B_0 \subset C_0 \subset A$, $\mu(C_0) > \alpha$, and $\mu(C) = \mu(C_0)$ for each $C \in \mathfrak{M}$ for which $B_0 \subset C \subset C_0$ and $\mu(C) > \alpha$. Conclude that $C_0 - B_0$ is an atom of μ.

(ii) If μ has no atoms and $\mu(X) = 1$, then for each $n = 1, 2, \ldots$ there are disjoint sets $A_i \in \mathfrak{M}$ such that $\mu(A_i) = 1/n$, $i = 1, \ldots, n$, and $\bigcup_{i=1}^{n} A_i = Z$. Hint: Use (i).

(18-29)*+ Let ν be a full probability in a set T. Prove that

(i) If ν has an atom, then there is a two-valued full probability μ in T. Hint: If $A \subset T$ is an atom of ν, let $\mu(B) = \nu(B \cap A)/\nu(A)$ for each $B \subset Z$.

(ii) If ν has no atoms, then there is a disjoint family $\mathfrak{D} \subset \exp T$ such that $\|\mathfrak{D}\| = \mathfrak{c}$, $\bigcup\{A : A \in \mathfrak{D}\} = T$, and $\nu(A) = 0$ for each $A \in \mathfrak{D}$. Hint: By (18-28)(ii), there are sets $A_{i,n} \subset T$, $i = 1, \ldots, n$, $n = 1, 2, \ldots$, such that $A_{i,n} \cap A_{j,n} = \emptyset$ whenever $i \neq j$, $\nu(A_{i,n}) = 1/n$, and $\bigcup_{i=1}^{n} A_{i,n} = T$. Define $\mathfrak{D}$ as a family of all sets $\bigcap_{n=1}^{\infty} A_{i_n,n}$, where i_n are integers and $1 \leq i_n \leq n$, $n = 1, 2, \ldots$.

(iii) If $\mathfrak{c}$ is not a measurable cardinal, then ν has an atom. Hint: Use (ii) and (18.21).

(iv) If $\mathfrak{c}$ is not a measurable cardinal, then any cardinal $\aleph$ is measurable iff it is two-valued measurable. Hint: Use (iii) and (i).

The remaining two exercises will show that the σ-additivity of a full probability is essential in defining measurable cardinals.

(18-30)$^{+*}$ Let T be a set. A family $\mathfrak{U} \subset \exp T$ is called a filter in T whenever

(a) $\emptyset \notin \mathfrak{U}$

(b) $A, B \in \mathfrak{U} \Rightarrow A \cap B \in \mathfrak{U}$

(c) $(A \in \mathfrak{U}, A \subset B \subset T) \Rightarrow B \in \mathfrak{U}$.

An ultrafilter in T is a filter in T which is not properly contained in any other filter in T. Thus an ultrafilter in T is a maximal element in the family of all filters in T which is partially ordered by inclusion. A filter $\mathfrak{U}$ in T is called free if $\bigcap\{A : A \in \mathfrak{U}\} = \emptyset$. Show that

(i) A filter $\mathfrak{U}$ in T is an ultrafilter in T iff for each $A \subset T$ either $A \in \mathfrak{U}$ or $T - A \in \mathfrak{U}$.

(ii) If T is infinite, then there is a free ultrafilter in T.

Hint: Observe that the complements of finite subsets of T form a free filter in T; apply Zorn's lemma.

(18-31)$^{+}$ Let T be an infinite set. Show that there is a finitely additive two-valued measure ν on exp T [see (8-5)] such that $\nu(T) = 1$ and $\nu(\{\alpha\}) = 0$ for each $\alpha \in T$. Hint: Using (18-30)(ii), choose a free ultrafilter $\mathfrak{U}$ in T, and for $A \subset T$, let $\nu(A) = 1$ if $A \in \mathfrak{U}$, and $\nu(A) = 0$ otherwise.

REFERENCES

1. Apostol, T. M., Mathematical Analysis, Addison-Wesley, Reading, 1964.
2. Arhangelskij, A. V., The power of bicompacta with first axiom of countability, Dokl. Akad. Nauk SSSR, 187(1969), 967-970; English translation: Soviet Math. Dokl., 10(1969), 951-955.
3. Bennett, H. R. and D. J. Lutzer, A note on weak θ-refinability, General Topology and Appl., 2(1972), 49-54.
4. Birkhoff, G., Lattice Theory, AMS, Providence, 1948.
5. Bourbaki, N., Intégration, Hermann, Paris, 1963.
6. Černý, I. and J. Mařík, The Integral Calculus I, Lecture notes of the Charles University, State Pedagogical Publishing Co., Prague, 1960.
7. Daniell, P., A general form of integral, Ann. of Math., 19(1917), 279-294.
8. Dugundji, J., Topology, Allyn and Bacon, Boston, 1966.
9. Gillman, L. and M. Jerison, Rings of Continuous Functions, Van Nostrand, New York, 1960.
10. Halmos, P. R., Measure Theory, Van Nostrand, New York, 1950.
11. Halmos, P. R., Naive Set Theory, Van Nostrand, New York, 1960.
12. Haydon, R., On compactness in spaces of measures and measure compact spaces, Proc. London Math. Soc., (3) 29(1974), 1-16.
13. Hewitt, E. and K. Stromberg, Real and Abstract Analysis, Springer, New York, 1965.
14. Jech, T. J., Lectures in Set Theory with Particular Emphasis on the Method of Forcing, Springer, New York, 1971.
15. Kakutani, S. and J. C. Oxtoby, Construction of a nonseparable invariant extension of the Lebesgue measure space, Ann. of Math., 52(1950), 580-590.
16. Kelley, J. L., General Topology, Van Nostrand, New York, 1955.

17. Loomis, L. H., An Introduction to Abstract Harmonic Analysis, Van Nostrand, New York, 1953.

18. Mařík, J., The Lebesgue integral in abstract spaces, Časopis pro pěst. mat., 76(1951), 175-194.

19. Natanson, I. P., Theory of Functions of a Real Variable, Ungar, New York, 1964.

20. Ross, K. A. and K. Stromberg, Baire sets and Baire measures, Arkiv fur Matematik, 6(1965), 151-160.

21. Solovay, R. M., Real-valued measurable cardinals, Axiomatic Set Theory, AMS, Providence, 1971, 397-428.

22. Stone, M. H., Notes on integration II, Proc. Nat. Acad. Sci., USA, 34(1948), 447-455.

23. Ungar, P. and M. Machover, Generating subsets of the plane, Amer. Math. Monthly, 82(1975), 677-678.

24. Wilbur, W. J., On measurability and regularity, Proceedings AMS, 21(1969), 741-746.

LIST OF FREQUENTLY USED SYMBOLS

INDEX